Understan... Series

Putting **Essential Understanding** of

Fractions
into Practice

in Grades
3–5

Kathryn Chval
University of Missouri
Columbia, Missouri

John Lannin
University of Missouri
Columbia, Missouri

Dusty Jones
Sam Houston State University
Huntsville, Texas

Kathryn Chval
Volume Editor
University of Missouri
Columbia, Missouri

Barbara J. Dougherty
Series Editor
University of Missouri
Columbia, Missouri

NATIONAL COUNCIL OF
TEACHERS OF MATHEMATICS

more4u
www.nctm.org/more4u
Access code: FRA14542

Copyright © 2013 by
The National Council of Teachers of Mathematics, Inc.
1906 Association Drive, Reston, VA 20191-1502
(703) 620-9840; (800) 235-7566; www.nctm.org

ISBN: 978-0-87353-732-2

The Cataloging-in-Publication Data are on file with the Library of Congress.

The National Council of Teachers of Mathematics is the public voice of mathematics education, providing vision, leadership, and professional development to support teachers in ensuring equitable mathematics learning of the highest quality for all students.

Printed in the United States of America

Contents

Chapter 3

Chapter 4

Chapter 5

Accompanying Materials at More4U

Foreword

Teaching mathematics in prekindergarten–grade 12 requires knowledge of mathematical content and developmentally appropriate pedagogical knowledge to provide students with experiences that help them learn mathematics with understanding, while they reason about and make sense of the ideas that they encounter.

In 2010 the National Council of Teachers of Mathematics (NCTM) published the first book in the Essential Understanding Series, focusing on topics that are critical to the mathematical development of students but often difficult to teach. Written to deepen teachers' understanding of key mathematical ideas and to examine those ideas in multiple ways, the Essential Understanding Series was designed to fill in gaps and extend teachers' understanding by providing a detailed survey of the big ideas and the essential understandings related to particular topics in mathematics.

The Putting Essential Understanding into Practice Series builds on the Essential Understanding Series by extending the focus to classroom practice. These books center on the pedagogical knowledge that teachers must have to help students master the big ideas and essential understandings at developmentally appropriate levels.

To help students develop deeper understanding, teachers must have skills that go beyond knowledge of content. The authors demonstrate that for teachers—

- understanding student misconceptions is critical and helps in planning instruction;

- knowing the mathematical content is not enough—understanding student learning and knowing different ways of teaching a topic are indispensable;

- constructing a task is important because the way in which a task is constructed can aid in mediating or negotiating student misconceptions by providing opportunities to identify those misconceptions and determine how to address them.

Through detailed analysis of samples of student work, emphasis on the need to understand student thinking, suggestions for follow-up tasks with the potential to move students forward, and ideas for assessment, the Putting Essential Understanding into Practice Series demonstrates best practice for developing students' understanding of mathematics.

The ideas and understandings that the Putting Essential Understanding into Practice Series highlight for student mastery are also embodied in the Common Core State

Standards for Mathematics, and connections with these new standards are noted throughout each book.

On behalf of the Board of Directors of NCTM, I offer sincere thanks to everyone who has helped to make this new series possible. Special thanks go to Barbara J. Dougherty for her leadership as series editor and to all the authors for their work on the Putting Essential Understanding into Practice Series. I join the project team in welcoming you to this special series and extending best wishes for your ongoing enjoyment—and for the continuing benefits for you and your students—as you explore Putting Essential Understanding into Practice!

Linda M. Gojak
President, 2012–2014
National Council of Teachers of Mathematics

Preface

The Putting Essential Understanding into Practice Series explores the teaching of mathematics topics in grades K–12 that are difficult to learn and to teach. Each volume in this series focuses on specific content from one volume in NCTM's Essential Understanding Series and links it to ways in which those ideas can be taught successfully in the classroom.

Thus, this series builds on the earlier series, which aimed to present the mathematics that teachers need to know and understand well to teach challenging topics successfully to their students. Each of the earlier books identified and examined the big ideas related to the topic, as well as the "essential understandings"—the associated smaller, and often more concrete, concepts that compose each big idea.

Taking the next step, the Putting Essential Understanding into Practice Series shifts the focus to the specialized pedagogical knowledge that teachers need to teach those big ideas and essential understandings effectively in their classrooms. The Introduction to each volume details the nature of the complex, substantive knowledge that is the focus of these books—*pedagogical content knowledge*. For the topics explored in these books, this knowledge is both student centered and focused on teaching mathematics through problem solving.

Each book then puts big ideas and essential understandings related to the topic under a high-powered teaching lens, showing in fine detail how they might be presented, developed, and assessed in the classroom. Specific tasks, classroom vignettes, and samples of student work serve to illustrate possible ways of introducing students to the ideas in ways that will enable students not only to make sense of them now but also to build on them in the future. Items for readers' reflection appear throughout and offer teachers additional opportunities for professional development.

The final chapter of each book looks at earlier and later instruction on the topic. A look back highlights effective teaching that lays the earlier foundations that students are expected to bring to the current grades, where they solidify and build on previous learning. A look ahead reveals how high-quality teaching can expand students' understanding when they move to more advanced levels.

Each volume in the Putting Essential Understanding into Practice Series also includes appendixes that list the big ideas and essential understandings related to the topic, detail resources for teachers, and present the tasks discussed in the book. These materials, which are available to readers both in the book and online at www.nctm.org/more4u, are intended to extend and enrich readers' experiences and possibilities for using the book. Readers can gain online access to these materials by going to the More4U website and entering the code that appears on the book's title page. They can then print out these materials for personal or classroom use.

Because the topics chosen for both the earlier Essential Understanding Series and this successor series represent areas of mathematics that are widely regarded as challenging to teach and to learn, we believe that these books fill a tangible need for teachers. We hope that as you move through the tasks and consider the associated classroom implementations, you will find a variety of ideas to support your teaching and your students' learning.

Acknowledgments from the Authors

We would like to thank the administrators and teachers at Paxton Keeley Elementary School, Grant Elementary School, Stewart Elementary School, and Huntsville Intermediate School for collaborating with us on the material for this volume. We would also like to thank their third-, fourth-, fifth-, and sixth-grade students who shared their mathematical thinking with us. In addition, we extend thanks to Chris Bowling for his assistance in creating figures and scanning students' work for this book.

Finally, Dusty Jones would like to thank the University of Florida, where he served as a visiting professor while editing later versions of the book.

Introduction

Shulman (1986, 1987) identified seven knowledge bases that influence teaching:

1. Content knowledge

2. General pedagogical knowledge

3. Curriculum knowledge

4. Knowledge of learners and their characteristics

5. Knowledge of educational contexts

6. Knowledge of educational ends, purposes, and values

7. Pedagogical content knowledge

The specialized content knowledge that you use to transform your understanding of mathematics content into ways of teaching is what Shulman identified as item 7 on this list—*pedagogical content knowledge* (Shulman 1986). This is the knowledge that is the focus of this book—and all the volumes in the Putting Essential Understanding into Practice Series.

Pedagogical Content Knowledge

In mathematics teaching, pedagogical content knowledge includes at least four indispensable components:

1. Knowledge of curriculum for mathematics

2. Knowledge of assessments for mathematics

3. Knowledge of instructional strategies for mathematics

4. Knowledge of student understanding of mathematics (Magnusson, Krajcik, and Borko 1999)

These four components are linked in significant ways to the content that you teach.

Even though it is important for you to consider how to structure lessons, deciding what group and class management techniques you will use, how you will allocate time, and what will be the general flow of the lesson, Shulman (1986) noted that it is even more important to consider *what* is taught and the *way* in which it is taught. Every day, you make at least five essential decisions as you determine—

1. which explanations to offer (or not);

2. which representations of the mathematics to use;

3. what types of questions to ask;

4. what depth to expect in responses from students to the questions posed; and

5. how to deal with students' misunderstandings when these become evident in their responses.

Your pedagogical content knowledge is the unique blending of your content expertise and your skill in pedagogy to create a knowledge base that allows you to make robust instructional decisions. Shulman (1986, p. 9) defined pedagogical content knowledge as "a second kind of content knowledge…, which goes beyond knowledge of the subject matter per se to the dimension of subject matter knowledge *for teaching.*" He explained further:

> Pedagogical content knowledge also includes an understanding of what makes the learning of specific topics easy or difficult: the conceptions and preconceptions that students of different ages and backgrounds bring with them to the learning of those most frequently taught topics and lessons. (p. 9)

If you consider the five decision areas identified at the top of the page, you will note that each of these requires knowledge of the mathematical content and the associated pedagogy. For example, an explanation that you give your students about comparing fractions requires that you take into account their knowledge of models of fractions, the relationships of fractions to other numbers such as whole numbers, and important ideas that you want students to understand about the magnitude of fractions. Your knowledge of fractions can help you craft tasks and questions that provide counterexamples and ways to guide your students in seeing connections across multiple number systems. As you establish the content, complete with learning goals, you then need to consider how to move your students from their initial understandings to deeper ones, building rich connections along the way.

The instructional sequence that you design to meet student learning goals has to take into consideration the misconceptions and misunderstandings that you might expect to encounter (along with the strategies that you expect to use to negotiate them), your expectation of the level of difficulty of the topic for your students, the progression of experiences in which your students will engage, appropriate collections of representations for the content, and relationships between and among fractions and other topics.

Model of Teacher Knowledge

Grossman (1990) extended Shulman's ideas to create a model of teacher knowledge with four domains (see fig. 0.1):

1. Subject-matter knowledge

2. General pedagogical knowledge

3. Pedagogical content knowledge

4. Knowledge of context

Subject-matter knowledge includes mathematical facts, concepts, rules, and relationships among concepts. Your understanding of the mathematics affects the way in which you teach the content—the ideas that you emphasize, the ones that you do not, particular algorithms that you use, and so on (Hill, Rowan, and Ball 2005).

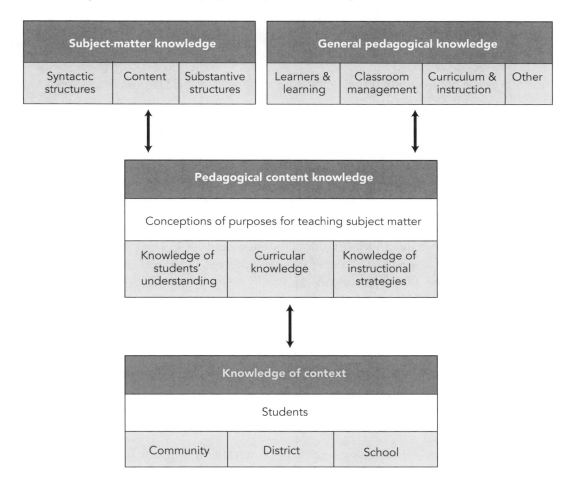

Fig. 0.1. Grossman's (1990, p. 5) model of teacher knowledge

Your pedagogical knowledge relates to the general knowledge, beliefs, and skills that you possess about instructional practices. These include specific instructional strategies that you use, the amount of wait time that you allow for students' responses to questions or tasks, classroom management techniques that you use for setting expectations and organizing students, and your grouping techniques, which might include having your students work individually or cooperatively or collaboratively, in groups or pairs. As Grossman's model indicates, your understanding and interpretation of the environment of your school, district, and community can also have an impact on the way in which you teach a topic.

Note that pedagogical content knowledge has four aspects, or components, in Grossman's (1990) model:

1. Conceptions of purposes for teaching

2. Knowledge of students' understanding

3. Knowledge of curriculum

4. Knowledge of instructional strategies

Each of these components has specific connections to the classroom. It is useful to consider each one in turn.

First, when you think about the goals that you want to establish for your instruction, you are focusing on your conceptions of the purposes for teaching. This is a broad category but an important one because the goals that you set will define learning outcomes for your students. These conceptions influence the other three components of pedagogial content knowledge. Hence, they appropriately occupy their overarching position in the model.

Second, your knowledge of your students' understanding of the mathematics content is central to good teaching. To know what your students understand, you must focus on both their conceptions and their misconceptions. As teachers, we all recognize that students develop naïve understandings that may or may not be immediately evident to us in their work or discourse. These can become deep-rooted misconceptions that are not simply errors that students make. Misconceptions may include incorrect generalizations that students have developed, such as thinking that the size of the denominator tells how "large" a fraction is. These generalizations may even be predictable notions that students exhibit as part of a developmental trajectory, such as thinking that doubling a fraction results in an equivalent fraction.

Part of your responsibility as a teacher is to present tasks or to ask questions that can bring misconceptions to the forefront. Once you become aware of misconceptions in students' thinking, you then have to determine the next instructional steps. The mathematical ideas presented in this volume focus on common misconceptions that students form in relation to a specific topic—fractions in grades 3–5. This book shows how the type of task selected and the sequencing of carefully developed questions can bring the misconceptions to light, as well as how particular teachers took the next instructional steps to challenge the students' misconceptions.

Third, curricular knowledge for mathematics includes multiple areas. Your teaching may be guided by a set of standards such as the Common Core State Standards for Mathematics (CCSSM; National Governors' Association Center for Best Practices and Council of Chief State School Officers 2010) or other provincial, state, or local standards. You may in fact use these standards as the learning outcomes for your students. Your textbook is another source that may influence your instruction. With any textbook also comes a particular philosophical view of mathematics, mathematics teaching, and student learning. Your awareness and understanding of the curricular perspectives related to the choice of standards and the selection of a textbook can help to determine how you actually enact your curriculum. Moreover, your district or school may have a pacing guide that influences your delivery of the curriculum. In this book, we can focus only on the alignment of the topics presented with broader curricular perspectives, such as CCSSM. However, your own understanding of and expertise with your other curricular resources, coupled with the parameters defined by the expected student outcomes from standards documents, can provide the specificity that you need for your classroom.

In addition to your day-to-day instructional decisions, you make daily decisions about which tasks from curricular materials you can use without adaptation, which tasks you will need to adapt, and which tasks you will need to create on your own. Once you select or develop meaningful, high-quality tasks and use them in your mathematics lesson, you have launched what Yinger (1988) called "a three-way conversation between teacher, student, and problem" (p. 86). This process is not simple—it is complex because how students respond to the problem or task is directly linked to your next instructional move. That means that you have to plan multiple instructional paths to choose among as students respond to those tasks.

Knowledge of the curriculum goes beyond the curricular materials that you use. You also consider the mathematical knowledge that students bring with them from grade 2 and what they should learn by the end of grade 5. The way in which you teach a foundational concept or skill has an impact on the way in which students will interact with and learn later related content. For example, the types of representations that you include in your introduction of fractions are the ones that your students will use to evaluate other representations and ideas in later grades.

Fourth, knowledge of instructional strategies is essential to pedagogical content knowledge. Having a wide array of instructional strategies for teaching mathematics is central to effective teaching and learning. Instructional strategies, along with knowledge of the curriculum, may include the selection of mathematical tasks, together with the way in which those tasks will be enacted in the classroom. Instructional strategies may also include the way in which the mathematical content will be structured for students. You may have very specific ways of thinking about how you will structure your presentation of a mathematical idea—not only how you will sequence the introduction and development of the idea, but also how you will present that idea to your students. Which examples should you select, and which questions should you ask? What representations should you use? Your knowledge of instructional strategies, coupled with your knowledge of your curriculum, permits you to align the selected mathematical tasks closely with the way in which your students perform those tasks in your classroom.

The instructional approach in this volume combines a student-centered perspective with an approach to mathematics through problem solving. A student-centered approach is characterized by a shared focus on student and teacher conversations, including interactions among students. Students who learn through such an approach are active in the learning process and develop ways of evaluating their own work and one another's in concert with the teacher's evaluation.

Teaching through problem solving makes tasks or problems the core of mathematics teaching and learning. The introduction to a new topic consists of a task that students work through, drawing on their previous knowledge while connecting it with new ideas. After students have explored the introductory task (or tasks), their consideration of solution methods, the uniqueness or multiplicity of solutions, and extensions of the task create rich opportunities for discussion and the development of specific mathematical concepts and skills.

By combining the two approaches, teachers create a dynamic, interactive, and engaging classroom environment for their students. This type of environment promotes the ability of students to demonstrate CCSSM's Standards for Mathematical Practice while learning the mathematics at a deep level.

The chapters that follow will show that instructional sequences embed all the characteristics of knowledge of instructional strategies that Grossman (1990) identifies. One component that is not explicit in Grossman's model but is included in a model developed by Magnusson, Krajcik, and Borko (1999) is the knowledge of assessment. Your knowledge of assessment in mathematics plays an important role in guiding your instructional decision-making process.

There are different types of assessments, each of which can influence the evidence that you collect as well as your view of what students know (or don't know) and how they know what they do. Your interpretation of what students know is also related to your view of what constitutes "knowing" in mathematics. As you examine the tasks, classroom vignettes, and samples of student work in this volume, you will notice that teacher questioning permits formative assessment that supplies information that spans both conceptual and procedural aspects of understanding. *Formative assessment,* as this book uses the term, refers to an appraisal that occurs during an instructional segment, with the aim of adjusting instruction to meet the needs of students more effectively (Popham 2006). Formative assessment does not always require a paper-and-pencil product but may include questions that you ask or tasks that students complete during class.

The information that you gain from student responses can provide you with feedback that guides the instructional flow, while giving you a sense of how deeply (or superficially) your students understand a particular idea—or whether they hold a misconception that is blocking their progress. As you monitor your students' development of rich understanding, you can continually compare their responses with your expectations and then adapt your instructional plans to accommodate their current levels of development. Wiliam (2007, p. 1054) described this interaction between teacher expectations and student performance in the following way:

> It is therefore about assessment functioning as a bridge between teaching and learning, helping teachers collect evidence about student achievement in order to adjust instruction to better meet student learning needs, in real time.

Wiliam notes that for teachers to get the best information about student understandings, they have to know how to facilitate substantive class discussions, choose tasks that include opportunities for students to demonstrate their learning, and employ robust and effective questioning strategies. From these strategies, you must then interpret student responses and scaffold their learning to help them progress to more complex ideas.

Characteristics of Tasks

The type of task that is presented to students is very important. Tasks that focus only on procedural aspects may not help students learn a mathematical idea deeply.

Superficial learning may result in students forgetting easily, requiring reteaching, and potentially affecting how they understand mathematical ideas that they encounter in the future. Thus, the tasks selected for inclusion in this volume emphasize deep learning of significant mathematical ideas. These rich, "high-quality" tasks have the power to create a foundation for more sophisticated ideas and support an understanding that goes beyond "how" to "why." Figure 0.2 identifies the characteristics of a high-quality task.

As you move through this volume, you will notice that it sequences tasks for each mathematical idea so that they provide a cohesive and connected approach to the identified concept. The tasks build on one another to ensure that each student's thinking becomes increasingly sophisticated, progressing from a novice's view of the content to a perspective that is closer to that of an expert. We hope that you will find the tasks useful in your own classes.

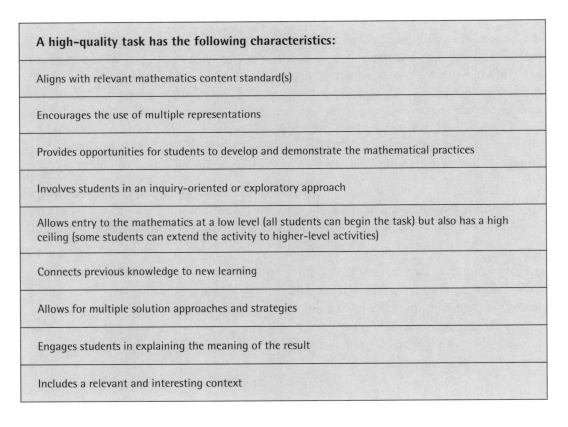

A high-quality task has the following characteristics:

Aligns with relevant mathematics content standard(s)

Encourages the use of multiple representations

Provides opportunities for students to develop and demonstrate the mathematical practices

Involves students in an inquiry-oriented or exploratory approach

Allows entry to the mathematics at a low level (all students can begin the task) but also has a high ceiling (some students can extend the activity to higher-level activities)

Connects previous knowledge to new learning

Allows for multiple solution approaches and strategies

Engages students in explaining the meaning of the result

Includes a relevant and interesting context

Fig. 0.2. Characteristics of a high-quality task

Types of Questions

The questions that you pose to your students in conjunction with a high-quality task may at times cause them to confront ideas that are at variance with or directly contradictory to their own beliefs. The state of mind that students then find themselves in is called *cognitive dissonance*, which is not a comfortable state for students—or, on occasion, for the teacher. The tasks in this book are structured in a way that forces students to deal with two conflicting ideas. However, it is through the process of negotiating the contradictions that students come to know the content much more deeply. How the teacher handles this negotiation determines student learning.

You can pose three types of questions to support your students' process of working with and sorting out conflicting ideas. These questions are characterized by their potential to encourage reversibility, flexibility, and generalization in students' thinking (Dougherty 2001). All three types of questions require more than a one-word or one-number answer. Reversibility questions are those that have the capacity to change the direction of students' thinking. They often give students the solution and require them to create the corresponding problem. A flexibility question can be one of two types: it can ask students to solve a problem in more than one way, or it can ask them to compare and contrast two or more problems or determine the relationship between or among concepts and skills. Generalization questions also come in two types: they ask students to look at multiple examples or cases and find a pattern or make observations, or they ask them to create a specific example of a rule, conjecture, or pattern. Figure 0.3 provides examples of reversibility, flexibility, and generalization questions related to fractions.

Type of question	Example
Reversibility question	What are three possible fractions that are greater than $\frac{1}{2}$ but less than $\frac{7}{8}$?
Flexibility question	How can you represent $\frac{1}{3} + \frac{1}{4}$ in two ways?
Flexibility question	How are fractions related to and different from whole numbers?
Generalization question	What patterns do you notice?
Generalization question	What are the characteristics of fractions that cluster around the benchmark fraction $\frac{1}{2}$?

Fig. 0.3. Examples of reversibility, flexibility, and generalization questions

Conclusion

The Introduction has provided a brief overview of the nature of—and necessity for—pedagogical content knowledge. This knowledge, which you use in your classroom every day, is the indispensable medium through which you transmit your understanding of the big ideas of the mathematics to your students. It determines your selection of appropriate, high-quality tasks and enables you to ask the types of questions that will not only move your students forward in their understanding but also allow you to determine the depth of that understanding.

The chapters that follow describe important ideas related to learners, curricular goals, instructional strategies, and assessment that can assist you in transforming your students' knowledge into formal mathematical ideas related to fractions. These chapters provide specific examples of mathematical tasks and student thinking for you to analyze to develop your pedagogical content knowledge for teaching fractions in grades 3–5 or to give you ideas to help other colleagues develop this knowledge. You will also see how to bring together and interweave your knowledge of learners, curriculum, instructional strategies, and assessment to support your students in grasping the big ideas and essential understandings and using them to build more sophisticated knowledge.

Students in grades 3–5 are involved in many situations outside their classrooms that affect their initial understanding of fractions. In addition, they have developed some ideas about fractions at earlier grade levels. Students in elementary classrooms frequently demonstrate understanding of mathematical ideas related to fractions in a particular context or in connection with a specific model or tool. Yet, in other situations these same students do not demonstrate that same understanding. As their teacher, you must understand the ideas that they have developed about fractions in their prior experiences so that you can extend this knowledge and see whether or how it differs from the formal mathematical knowledge that they need to be successful in reasoning with fractions. You have the important responsibility of assessing their current knowledge related to the big ideas of fractions as well as their understanding of various representations of fractions and their power and limitations. Your understanding will facilitate and reinforce your instructional decisions. Teaching the big mathematical ideas and helping students develop essential understandings related to fractions is obviously a very challenging and complex task.

practice

Chapter 1
From Whole Numbers to Fractions

Essential Understanding 1c

The rational numbers allow us to solve problems that are not possible to solve with just whole numbers or integers.

Big Idea 1 for rational numbers, as stated in *Developing Essential Understanding of Rational Numbers for Teaching Mathematics in Grades 3–5* (Barnett-Clarke et al. 2010), captures the foundational notion that extending from whole numbers to rational numbers creates a more powerful and complicated number system. This system allows students to solve problems that they cannot solve with just whole numbers or integers. This is the important insight that students gain when they grasp Essential Understanding 1c, the idea on which this chapter focuses.

Working toward Essential Understanding 1c

Designing, adapting, or selecting worthwhile assessment tasks, interpreting the responses of individual students, and making instructional decisions based on the results requires specialized knowledge. For example, Reflect 1.1 asks you to consider the mathematical assessment tasks shown in figures 1.1 and 1.2. As you examine these tasks, which were used with students entering grades 3–5, think about the questions for reflection.

Reflect 1.1

What aspects of students' understanding of fractions are being assessed in figures 1.1 and 1.2?

Do you predict that students in grades 3–5 would perform better on the task in figure 1.1 or on the task in figure 1.2? Why?

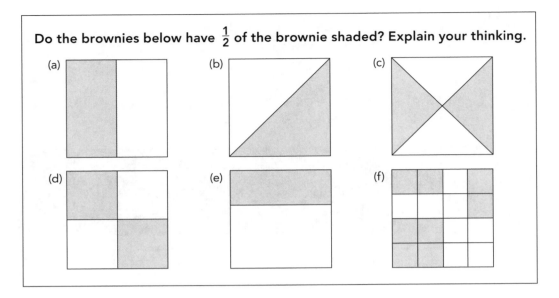

Fig. 1.1. A fraction task exploring the meaning of one-half

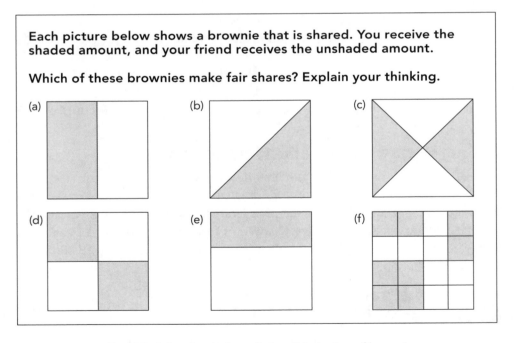

Fig. 1.2. A fraction task exploring the sharing of brownies

The mathematical tasks in figures 1.1 and 1.2 are closely connected, but they also have an important distinction. The task in figure 1.1 requires students to determine whether the shading represents one-half, whereas the task in figure 1.2 involves determining whether the diagrams are equally partitioned, or "fair shared." Battista (2012) argues that a student may understand partitioning but have no concept of the meaning of fractions. To some extent, these two tasks assess similar ideas about the meaning that students give to one-half. In particular, for both tasks, students must consider whether the shaded and nonshaded regions are the same size.

Many times, your students' work can provide you with further insight into their understanding of the mathematical ideas that emerge in a task, allowing you greater insight into their understandings and misconceptions. Consider the work of two students on these two tasks. Figures 1.3 and 1.4 present the work of Miriam, a student who is entering grade 4. Figures 1.5 and 1.6 show the work of Jaden, a student who is entering grade 5. As you look at Miriam's and Jaden's work, consider what it reveals about their understanding or misunderstanding of fractions. Miriam and Jaden completed the tasks in figures 1.1 and 1.2 during the same class period. Moreover, their work is representative of the way in which the majority of children to whom we provided these items responded in grades 3–5.

For the task shown in figure 1.1, Miriam decided that only parts (a) and (b) have $\frac{1}{2}$ of the brownie shaded. For part (c), she stated, "No. Because half is 2 peices and this is 4 peices" (spelling as in original). She provided a similar argument for parts (d) and (f). She claimed that part (e) does not have $\frac{1}{2}$ shaded, "because half is 2 equals peices and these are not equals." Jaden provided similar responses for this task. On part (f), he wrote, "No. Because there are 16 in all and 8 shaded there has to be 1 and 2." He also provided an example that looked like the picture in (a).

Both Miriam and Jaden appear to have what Tall and Vinner (1981) refer to as a "concept image" of $\frac{1}{2}$ as a whole divided into two equal pieces, with one of the parts shaded. A concept image includes the mental images and associated properties and actions. Tall and Vinner explain that a concept image builds over years and changes as students mature and encounter conflicting ideas. Miriam and Jaden have a mental image of $\frac{1}{2}$ that is likely to have developed in their out-of-school experiences with the word "half"—experiences in which "half" is often used to mean a partition into two equal parts. When they think about the representation of the fraction $\frac{1}{2}$, they suppose that the unit must be divided into two, and only two, congruent pieces.

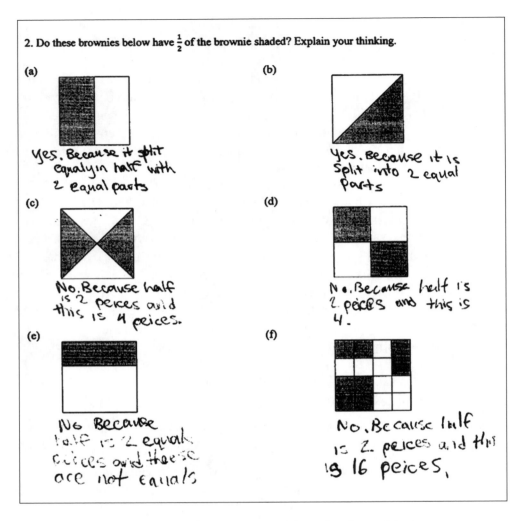

2. Do these brownies below have $\frac{1}{2}$ of the brownie shaded? Explain your thinking.

(a)

yes. Because it split
equally in half with
2 equal parts

(b)

yes. Because it is
Split into 2 equal
Parts

(c)

No. Because half
is 2 peices and
this is 4 peices.

(d)

No. Because half is
2 peices and this is
4.

(e)

No Because
half is 2 equal
peices and these
are not equals

(f)

No. Because half
is 2 peices and this
is 16 peices.

Fig. 1.3. Miriam's responses to the task in figure 1.1

For the task in figure 1.2, Miriam stated that all of the brownies were divided into fair shares except for part (e). For example, when evaluating the partitioning of the brownie in part (c), she noted, "This can [be fair shares] because it has 4 equal

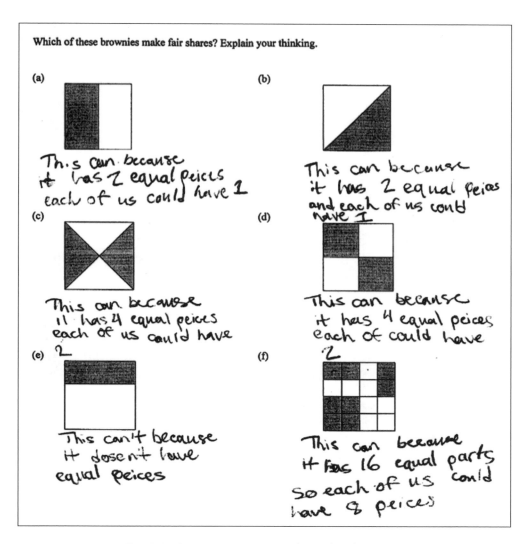

Which of these brownies make fair shares? Explain your thinking.

(a)

Th.s can because it has 2 equal peices each of us could have 1

(b)

This can because it has 2 equal peies and each of us could have I

(c)

This can because il has 4 equal peices each of us could have 2

(d)

This can because it has 4 equal peices each of could have 2

(e)

This can't because It dosen't have equal peices

(f)

This can because it has 16 equal parts so each of us could have 8 peices

Fig. 1.4. Miriam's responses to the task in figure 1.2

peices each of us could have 2." In the same way, Jaden wrote for part (f), "Fair. We each get 8 pieces." Compare Miriam's and Jaden's responses and then respond to the questions in Reflect 1.2.

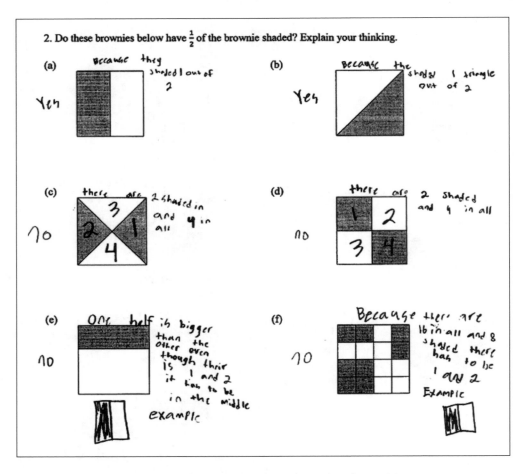

2. Do these brownies below have $\frac{1}{2}$ of the brownie shaded? Explain your thinking.

(a) Yes
Because they shaded 1 out of 2

(b) Yes
Because the shads 1 triangle out of 2

(c) No
there are 2 shaded in and 4 in all
3 2 1 4

(d) No
there are 2 shaded and 4 in all
1 2 3 4

(e) No
One half is bigger than the other even though their is 1 and 2 it has to be in the middle
example

(f) No
Because there are 16 in all and 8 shaded there has to be 1 and 2 Example

Fig. 1.5. Jaden's responses to the task in figure 1.1

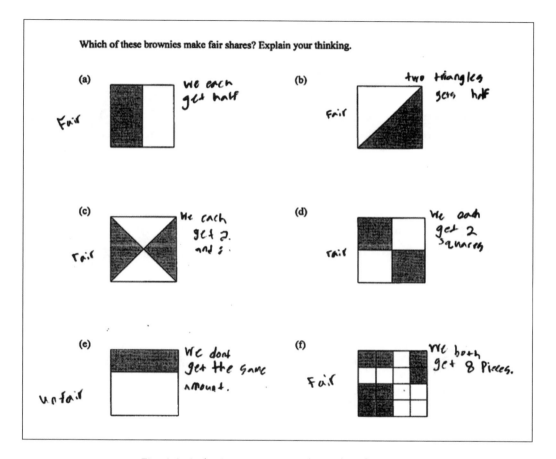

Which of these brownies make fair shares? Explain your thinking.

(a) Fair — we each get half

(b) Fair — two triangles gets half

(c) Fair — We each get 2. and ...

(d) Fair — We each get 2 squares

(e) unfair — We dont get the same amount.

(f) Fair — we both get 8 pieces.

Fig. 1.6. Jaden's responses to the task in figure 1.2

Reflect 1.2

What do Miriam and Jaden appear to understand about fair shares for two people?

How do their responses on the various parts of the task in figure 1.2 conflict with their responses on the parallel parts of the task in figure 1.1?

Both Miriam and Jaden recognize that making "fair shares" for two people does not require that the whole be split into only two pieces. They demonstrate that they have a strong understanding of "fair sharing" between two people that is disconnected from their understanding of the formal mathematical meaning of $\frac{1}{2}$.

You may wonder whether Miriam and Jaden recognize that the fractions $\frac{1}{2}$, $\frac{2}{4}$, and $\frac{8}{16}$ are equivalent. Another task on the same assessment asked students to list all the fractions they could that are equivalent to $\frac{1}{2}$. Miriam listed $\frac{2}{4}$, $\frac{4}{8}$, $\frac{8}{16}$, $\frac{16}{32}$, and $\frac{32}{64}$, and Jaden listed $\frac{2}{4}$, $\frac{1}{2}$, $\frac{3}{6}$, $\frac{4}{8}$, $\frac{5}{10}$, $\frac{7}{14}$, $\frac{16}{32}$, and $\frac{118}{236}$. It is clear that both students are familiar with a procedure for writing equivalent fractions, but their prior knowledge and previously developed meaning for $\frac{1}{2}$ interfered with their ability to make connections between the symbols and diagrams that represent $\frac{1}{2}$.

It is critical not only to consider students' mathematical conceptions or misconceptions, but also to provide them with possibilities for developing their understanding further. Consider how you might extend the understanding of $\frac{1}{2}$ that Miriam and Jaden currently demonstrate. It appears that they both understand the idea of fair sharing between two people, and they also have an idea of representations of equivalent fractions. One strategy might be to pose one or more of the questions shown in figure 1.7.

Examining how Miriam and Jaden responded to the tasks in figure 1.7 would allow further assessment of their mathematical understanding and suggest ways of extending it. You might anticipate that Miriam and Jaden would respond that the first two brownies have $\frac{2}{4}$ shaded, and the third has $\frac{8}{16}$ shaded. If they used their knowledge that $\frac{2}{4}$ and $\frac{8}{16}$ are equivalent to $\frac{1}{2}$, they might realize that the three models do have $\frac{1}{2}$ shaded. Furthermore, part (b) of the task in figure 1.7 could

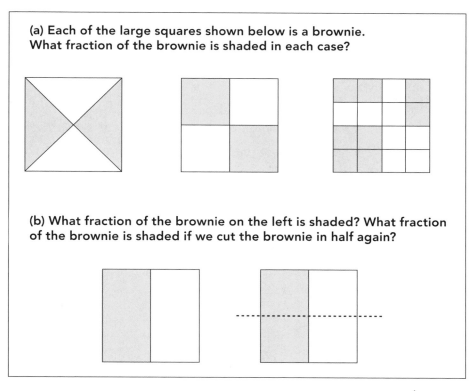

(a) Each of the large squares shown below is a brownie. What fraction of the brownie is shaded in each case?

(b) What fraction of the brownie on the left is shaded? What fraction of the brownie is shaded if we cut the brownie in half again?

Fig. 1.7. Questions to extend Miriam and Jaden's understanding of $\frac{1}{2}$

formalize this insight for them. They should recognize that the brownie on the left has $\frac{1}{2}$ shaded. By introducing the idea of partitioning the brownie again, this part of the task could help Miriam and Jaden recognize that the size of the shaded region did not change as a result.

Further, Miriam and Jaden appear to need assistance in connecting their informal ideas about sharing with the meaning of $\frac{1}{2}$. Discussing the students' various responses to the tasks in figures 1.1 and 1.2 in a whole-class session might allow you to make the idea explicit that when shapes are divided into two equal-sized shares, each share represents $\frac{1}{2}$, although the two shares do not have to be the same shape, adjacent to each other, or "split" with just one cut, as is the brownie on the left in part (b) of the task in figure 1.7.

Next, consider the work of Henri, a student who has just completed the fourth grade. In his response to the task in figure 1.1, Henri stated that only (a) and (b) show brownies with $\frac{1}{2}$ of the brownie shaded. He wrote, "The answer is a or b because $\frac{1}{2}$ of a something is like you have candy and you give one to yourself and to your friend. That's $\frac{1}{2}$." He said that the shaded portions of the brownies in part (c) and part (d) did not represent $\frac{1}{2}$, offering the following explanation: "No, it is not a $\frac{1}{2}$ because they are separated. But look at a and b. They have to be together like a or b." In evaluating the shaded portions of the brownie in part (f), he wrote, "They're not $\frac{1}{2}$ because look at f. It doesn't even look close to be a $\frac{1}{2}$." In completing the task in figure 1.2, Henri stated that all the shares of brownies were fair except for those shown in part (e).

Questions to gather information from Henri
1. Why do the shaded pieces have to touch?

2. Is $\frac{1}{2}$ of the brownie shaded in the following diagrams? Why or why not?

Questions to help move Henri forward
3. If you rearranged the shaded pieces in the figures above, would they still represent $\frac{1}{2}$? Why or why not?

4. John looked at the brownie on the left and said, "This has $\frac{1}{2}$ shaded. If you flip one of the shaded pieces over, it adds to half." Do you agree with John? Use John's strategy to determine if the other brownies have $\frac{1}{2}$ shaded.

Fig. 1.8. Questions for Henri

What does Henri appear to understand about the meaning of $\frac{1}{2}$? Clearly, something is missing, given his claim that the pieces have to be "together." Using a questioning strategy may be helpful. You can use questions to gather information about what your students understand. Moreover, effective questioning can help move them forward. Think of some questions that would give you more information about what Henri understands about $\frac{1}{2}$, along with some questions that would help him gain a better understanding of $\frac{1}{2}$. Figure 1.8 presents several such questions.

John, who has completed the fourth grade, stated that all the diagrams in figure 1.1 show $\frac{1}{2}$ except for part (e). In response to part (c), he wrote, "Yes, if you flip one of the shaded pieces over it adds to half." He used similar reasoning on part (f) and indicated that the two shaded squares in the upper right corner could fill in the gap on the left.

It may be that the concept images that Henri and John hold for $\frac{1}{2}$ allow a figure to be partitioned into more than two pieces, but Henri's requires that the shaded pieces be touching. John's, by contrast, permits mental rearrangement of the pieces and accepts that the shaded pieces may be separated from one another.

From what John has stated, it is not clear whether or not he believes that the pieces need to be the same shape. That is to say that as his teacher, you would not necessarily know how he might complete the task in figure 1.9, which shows a diagram in which the shaded pieces do not fit conveniently together.

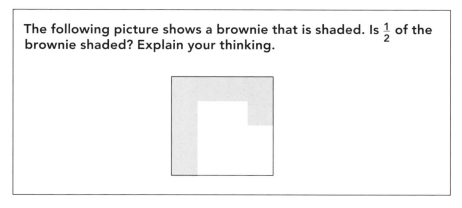

The following picture shows a brownie that is shaded. Is $\frac{1}{2}$ of the brownie shaded? Explain your thinking.

Fig. 1.9. An extension of task in figure 1.1

Figures 1.10 and 1.11 show assessment tasks for students entering grades 4–6. Reflect 1.3 poses questions for you to consider as you examine and compare these tasks with the previous ones in figures 1.1 and 1.2.

Reflect 1.3

Examine the tasks for students entering grades 4–6 shown in figures 1.10 and 1.11. What aspects of students' understanding of fractions are being assessed in these two tasks?

Do you think that students who have completed grades 3–5 are likely to perform better on the task in figure 1.10 or on that in figure 1.11? Why?

Suppose that a student in your class said the following:

$\frac{1}{4}$ is more than $\frac{1}{3}$ because 4 is more than 3.

Do you agree with this student? Explain your thinking.

Fig. 1.10. A fraction task about $\frac{1}{4}$ and $\frac{1}{3}$

Look at the shaded part of the two brownies below. Circle the brownie that has a larger amount of a brownie shaded.

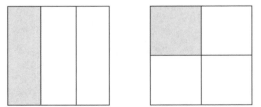

Explain why you think this is a larger amount of brownie.

Fig. 1.11. A fraction task about a larger amount of brownie

The mathematical tasks in figures 1.10 and 1.11 are closely connected, but they also have an important distinction. The first task asks students to evaluate another student's reasoning that is based on a common misconception. The second task, in contrast, requires students to determine which diagram displays a larger shaded portion of the whole. To an extent, these two tasks assess similar ideas about comparing fractions. Whereas the first task focuses students on the symbolic forms $\frac{1}{4}$ and $\frac{1}{3}$, the second uses diagrams of $\frac{1}{4}$ and $\frac{1}{3}$.

We provided these tasks to students entering grades 4–6 during the same class period. On one hand, in response to the task in figure 1.10, some students agreed while others disagreed, for various reasons. On the other hand, almost every student circled the brownie on the left in the task in figure 1.11, signifying that a larger amount was shaded. However, the reasons that students gave for this choice differed. Consider, for example, work on these tasks by Ryan (entering fourth grade) and Amber (entering sixth grade) and what it reveals about their understanding or misunderstanding of fractions. Ryan's work is shown in figures 1.12 and 1.13.

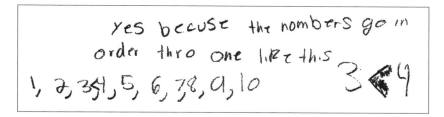

Fig. 1.12. Ryan's response to the task in figure 1.10

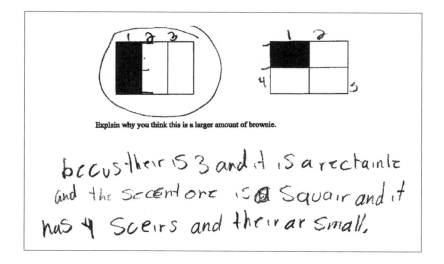

Explain why you think this is a larger amount of brownie.

Fig. 1.13. Ryan's response to the task in figure 1.11

As Ryan's teacher, you could ask Ryan to write a fraction for the shaded pieces of brownies shown in figure 1.11. This could help him connect the symbols $\frac{1}{4}$ and $\frac{1}{3}$ to the pictures used in the tasks, as well as highlight the role of the denominator in a fraction. The connection between pictorial and symbolic representation could also emphasize that the ordering for whole numbers does not directly apply to the case presented in figure 1.10.

Amber's responses to both of these tasks are shown in figures 1.14 and 1.15. Use the questions in Reflect 1.4 to guide you in assessing her understanding and misunderstanding.

Reflect 1.4

What does Amber appear to understand about the fractions $\frac{1}{3}$ and $\frac{1}{4}$?

How does her understanding compare with Ryan's?

What questions could you ask Amber that might help her clarify or strengthen her understanding of the meaning of $\frac{1}{3}$ and $\frac{1}{4}$?

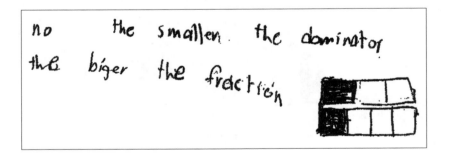

Fig. 1.14. Amber's responses to the task in figure 1.10

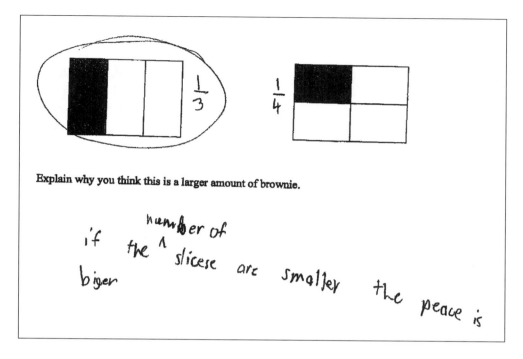

Fig. 1.15. Amber's responses to the task in figure 1.11

Amber's work indicates that Amber understands the meaning of both the pictorial and symbolic forms of $\frac{1}{3}$ and $\frac{1}{4}$ and demonstrates that she has made connections between the two forms. Without prompting, in her response to the task in figure 1.10, she draws a pictorial model of $\frac{1}{4}$ and $\frac{1}{3}$ to explain why $\frac{1}{4}$ is smaller than $\frac{1}{3}$ (see fig. 1.14). In her response to the task in figure 1.11, she correctly labels the models (see fig. 1.15).

In response to the task in figure 1.10, Amber wrote, "the smaller the d[en]ominator, the big[g]er the fraction." This reasoning works well here, because the numerators are the same. But what would you expect if you asked Amber to compare fractions with different numerators, such as $\frac{2}{3}$ and $\frac{3}{4}$, or different denominators, like $\frac{5}{8}$ and $\frac{3}{5}$? Amber would probably realize that $\frac{5}{8}$ was made of smaller pieces, but her response to the task in figure 1.10 does not make clear how she would use the numerator to answer this question.

Summarizing Pedagogical Content Knowledge to Support Essential Understanding 1c

Teaching the mathematical ideas in this chapter requires specialized knowledge related to the four components presented in the Introduction: learners, curriculum, instructional strategies, and assessment. The four sections that follow summarize some examples of these specialized knowledge bases in relation to Essential Understanding 1c. Although we separate them to highlight their importance, we also recognize that they are connected and support one another.

Knowledge of learners

Battista (2012, p. 1) states an important idea that has been illustrated by the students' work included in this chapter:

> Before students can understand fractions, they must understand partitioning. To partition a whole is to divide it into equal portions, like dividing a pizza equally among four people. Being able to partition, however, does not mean that one understands fractions.

Battista goes on to outline eight levels of sophistication in students' reasoning about fractions. At level 0, students may understand equal partitioning but not yet have an understanding of fractions. At level 1, students recognize only familiar diagrams of fractions, whereas at level 2, they understand fractions as numbers that involve counting all parts and shaded parts. As evidenced by their work on the task in figure 1.2 (see figs. 1.4 and 1.6), Miriam and Jaden appear to have a strong understanding of partitioning or fair shares, and, as their work on the later assessment item demonstrates, they can identify some examples of fractions that are equivalent to $\frac{1}{2}$. They also demonstrated their ability to recognize common images of $\frac{1}{2}$ in their responses to parts (a) and (b) in figure 1.1. Yet, they share a misconception that $\frac{1}{2}$ can be depicted only in diagrams that include two pieces that are congruent. It is critical that as a teacher you develop knowledge specific to learners' mathematical understandings and misconceptions to design instruction, assessment, and curricular tasks that will enhance their understanding.

Knowledge of curriculum

Barnett-Clarke and colleagues (2010, p. 19) discuss the importance of recognizing three features of rational numbers to understand them:

Rational numbers can be used in many different contexts, their interpretation can change depending on the context, and defining the unit is key to the interpretation (Behr et al. 1983; Carraher 1992, 1996; Kieren 1992; Lamon 2007).

These three ideas are crucial for students to know and understand, and consequently they should influence the selection of curricular tasks that use different contexts (for example, brownies, people, string), different language (the tasks in figs. 1.1 and 1.2, for instance, use the same diagrams but introduce the task in different words), and different units. As you think about which contexts, language, and units to use to explore and extend your students' mathematical ideas, it is also important to consider which ones are effective for which specific purposes. Fosnot and Dolk (2002, p. 28) argue persuasively for the use of meaningful contexts:

> When the context is a good one, the children talk about the situation. When a problem is camouflaged school mathematics, children talk about numbers abstractly; they lose sight of the problem as they try to figure out what the teacher wants.

You also need to carefully select tasks that will not perpetuate misconceptions, such as any faulty concept images that your students may have. For example, you may have students who assume that parts must be congruent (that is, the same size and shape) on the basis of the examples that they have experienced. Students should recognize that parts may be the same size without being the same shape.

Knowledge of instructional strategies

Teachers have a multitude of instructional strategies to draw on in developing students' understanding of fractions. The preceding discussion has suggested a few examples, which this section highlights.

Empson (1995) writes, "Most young children have experienced equal sharing and have quite a bit of intuitive knowledge of those situations" (p. 110). Capitalizing on those prior experiences and knowledge is an important instructional strategy. Empson also argues that the context of equal sharing can help children understand that pieces need to be the same size. The cases of Miriam and Jaden, whose performance on the brownie sharing task in figure 1.2 was stronger than on the more abstract task in figure 1.1, illustrate the truth of these observations.

A second strategy is to ask students to interpret diagrams such as those in the task in figure 1.1, but, more important, to create tasks that will encourage comparisons and connections. Fosnot and Dolk (2002) suggest that students should wonder, notice patterns, and ask questions such as "Why?" and "What if?" Facilitating a whole-class discussion in which your students compare and contrast tasks like those in figures 1.1 and 1.2 will generate interesting ideas and questions from them. This strategy will help students connect their knowledge about sharing with the meanings of fractions.

Another instructional strategy illustrated in this chapter's examples is to engage students in analyzing responses from fictitious students to a particular task. The task in figure 1.10 illustrates this approach: students must agree or disagree with the conjecture from a fictitious student in their class. Again, this strategy can lead to interesting whole-class discussions. In this case, the fictitious student's conjecture involves an erroneous connection between whole numbers and fractions, implying that comparing fractions is similar to comparing whole numbers. This is a misconception that teachers in grades 3–5 frequently encounter among students. Rather than introducing this misconception through the work of an actual student in the classroom, the "fictitious student" strategy permits highlighting this idea and initiating a discussion through the use of the work of "other students." Students can then justify their reasoning for their agreement or disagreement.

Knowledge of assessment

As noted in the Introduction, Wiliam (2007) emphasizes the importance of selecting tasks that provide opportunities for students to demonstrate their thinking and for teachers to make instructional decisions. The work that students generated in response to the tasks presented in this chapter provides evidence about their understanding and misunderstanding of fractions. The tasks ask students to explain their thinking or justify a response to give their teachers more information to use in interpreting their thinking and making determinations about future instruction.

Another assessment strategy that the examples in this chapter have illustrated is the questioning that can and should follow the initial assessment. Miriam's and Jaden's teacher made some decisions about what questions and tasks to pose next. This strategy may serve the purpose of enabling the teacher to collect additional assessment information, positioning the teacher to move the students' understanding forward or to challenge an identified misconception.

Conclusion

The children's work highlighted in this chapter has illustrated knowledge that teachers need to design, adapt, and select worthwhile curricular and assessment tasks related to fractions; interpret responses from specific students; and make instructional decisions based on the results. The chapter has discussed some essential understandings that children need to develop—as well as misconceptions that teachers need to challenge—to enable them to be successful with fractions. As teachers, we need to assess and build on children's understanding of partitioning, fair shares, and the meaning of fractions. Further, we need to determine whether our students have any faulty concept images of fractions. This work requires not only careful selection of tasks but also effective questions, as discussed in the Introduction. This effort will help students lay a foundation for developing the critical understanding of unit that we discuss in the next chapter.

practice

Chapter 2
The Concept of Unit

Essential Understanding 2*a*

The concept of *unit* is fundamental to the interpretation of rational numbers.

Helping students identify the unit, or whole, and connect it with a fractional part is an essential part of developing and deepening their understanding of the meaning of fractions, as well as laying the foundation for their future computational work with fractions. The idea of unit is presented as Essential Understanding 2*a* in *Developing Essential Understanding of Rational Numbers for Teaching Mathematics in Grades 3–5* (Barnett-Clarke et al. 2010), and understanding a fraction in relation to the unit appears as the first standard for fractions for grade 3 in the Common Core State Standards for Mathematics (CCSSM; National Governors Association Center for Best Practices and Council of Chief State School Officers 2010):

> Understand a fraction $\frac{1}{b}$ as the quantity formed by 1 part when a whole is partitioned into b equal parts; understand a fraction $\frac{a}{b}$ as the quantity formed by a parts of size $\frac{1}{b}$. (3.NF.1, p. 24)

Developing students' understanding of the importance and meaning of the unit in a fraction is the focus of this chapter.

Working toward Essential Understanding 2*a*

As the discussion of Essential Understanding 1*c* in Chapter 1 demonstrated, students have different intuitive understandings of fraction concepts. Your goal as a teacher is to formalize these ideas. As your students work with whole numbers and fractions outside of school, they naturally relate number to quantities, and such quantities are assigned units. For example, no one refers to having "2" but instead might refer to having "2 baseballs" or "2 apples" or being "2 years old." In a similar way, units are used with fractions that students encounter in their experiences outside the classroom. Signs say that the next interstate exit is $\frac{1}{2}$ mile ahead, a recipe requires $\frac{3}{4}$ of a cup of flour, or someone states that a particular child is $2\frac{1}{2}$ years old. A unit is provided that establishes meaning for the fractional amount. For example, saying that someone is $2\frac{1}{2}$, in reference to age, would lead to questions about whether this person is $2\frac{1}{2}$ years old, $2\frac{1}{2}$ weeks old, or $2\frac{1}{2}$ months old, to name a few possible units. Similarly, when introducing students to fractions in the classroom, it is important to clarify the unit that is referred to in the situation and to build on units that students have some familiarity with from their in- and out-of-school experiences.

The work of three fictitious students in relation to a given unit, a brownie, appears in the task shown in figure 2.1. Sally, Marcus, and Demetrius respond to a problem that shows a whole brownie and then asks what part is shaded in a representation of two partitioned brownies. Consider the questions in Reflect 2.1 as you explore these fictitious students' responses, which show different ways of thinking about the representation and its relationship to the identified unit.

Reflect 2.1

What aspects of students' understanding of fractions are being assessed in the task shown in figure 2.1? How would you expect students in grade 3–5 to respond to this task?

Which of the three students—Sally, Marcus, or Demetrius—provided a correct fraction name and referred to the correct corresponding unit?

This mathematical task is designed to assess how students view the unit in relation to various fraction names, such as "$\frac{3}{2}$ of a brownie" or "$\frac{3}{4}$ of 2 brownies." Students who accept Sally's fraction name and explanation apparently view the unit as as 2 brownies, and those who accept Marcus's or Demetrius's name for the fraction, along with the corresponding explanation, view the unit as a single brownie. Although the primary focus of the task is students' views of the unit, the students

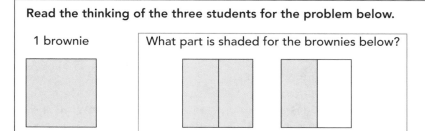

Read the thinking of the three students for the problem below.

1 brownie

What part is shaded for the brownies below?

(a) *Sally:* I think $\frac{3}{4}$ of the 2 brownies is shaded. The brownies are cut into 4 equal parts and 3 are shaded.

Is Sally correct? Explain your thinking.

(b) *Marcus:* I think $1\frac{1}{2}$ brownies are shaded. One of the brownies is shaded and $\frac{1}{2}$ of another brownie is shaded, so $1\frac{1}{2}$ brownies are shaded.

Is Marcus correct? Explain your thinking.

(c) *Demetrius:* I think that $\frac{3}{2}$ of a brownie is shaded. Each brownie is cut in half and 3 of the halves are shaded.

Is Demetrius correct? Explain your thinking.

Fig. 2.1. A task in which fictitious students assign different fraction names to the shaded part of the brownies

need to recognize that $\frac{3}{4}$ means 3 parts, each of which is $\frac{1}{4}$ of the unit, and that $\frac{3}{2}$ refers to 3 parts, each of which is $\frac{1}{2}$ of the unit.

All three fictitious students provide a correct fraction name and corresponding unit. Sally views both brownies as the unit, referring to the shaded part as $\frac{3}{4}$ of 2 brownies. Because the 2 brownies are divided into 4 equal-sized parts and 3 parts are shaded, the diagram represents $\frac{3}{4}$ of the 2 brownies. Marcus refers to a single brownie as the unit, as indicated by the label "brownies" at the end of his statement about $1\frac{1}{2}$ brownies. Because one brownie is shaded and $\frac{1}{2}$ of the other brownie is shaded, he properly identifies the shaded region as "$1\frac{1}{2}$ brownies." Demetrius refers to the same single-brownie unit as Marcus and notes that the shaded amount is $\frac{3}{2}$ of a brownie. Because each brownie is partitioned into 2 equal-sized parts, with each part making $\frac{1}{2}$ of a brownie, and 3 of the halves are shaded, Marcus reasons that the shaded region represents $\frac{3}{2}$ of a brownie. This task highlights that it is important for students to recognize that the unit can be viewed in different ways—as 2 brownies or as 1 brownie, for instance.

The chart in figure 2.2 provides an overview of how students in grades 4–6 responded to the work of the three fictitious students on the problem in figure 2.1. In each grade, at least half of the students wrote that Marcus was correct. Note that far fewer students agreed with Sally or Demetrius. Only 4 of 52 students, or not quite 8 percent, claimed that all three fictitious students were correct.

	Sally is correct	Marcus is correct	Demetrius is correct
Fourth grade (n = 16)	31% ($\frac{5}{16}$)	50% ($\frac{8}{16}$)	31% ($\frac{5}{16}$)
Fifth grade (n = 19)	47% ($\frac{9}{19}$)	84% ($\frac{16}{19}$)	32% ($\frac{6}{19}$)
Sixth grade (n = 17)	24% ($\frac{4}{17}$)	88% ($\frac{15}{17}$)	41% ($\frac{7}{17}$)

Fig. 2.2. Students' responses to Sally's, Marcus's, and Demetrius's assigning of different fraction names for two brownies

Closer examination of the student work on this task affords more insight into the mathematical thinking that teachers can reveal by using a task that allows students to analyze different views of unit. Figures 2.3, 2.4, and 2.5 show, respectively, the work of three students—Rafael (entering grade 6), Andy (entering grade 6), and David (entering grade 5)—in response to Sally's suggestion that $\frac{3}{4}$ of 2 brownies is shaded. As you study these three responses, consider the questions in Reflect 2.2, which focus attention on the mathematical understandings and misunderstandings that these students exhibit and specific strategies for moving them forward.

Reflect 2.2

How would you characterize the mathematical understandings and misunderstandings that Rafael, Andy, and David exhibit in their work, shown in figures 2.3, 2.4, and 2.5, respectively?

What specific strategies or questions would you use to move these students forward?

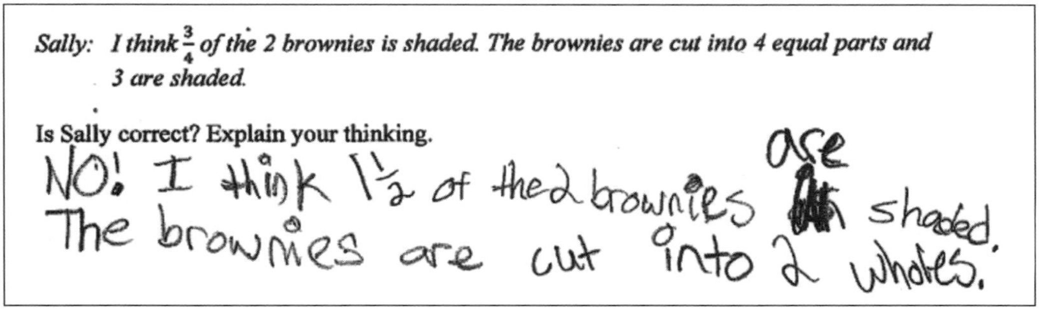

Sally: *I think $\frac{3}{4}$ of the 2 brownies is shaded. The brownies are cut into 4 equal parts and 3 are shaded.*

Is Sally correct? Explain your thinking.

NO! I think 1½ of the 2 brownies are shaded. The brownies are cut into 2 wholes.

Fig. 2.3. Rafael's response to Sally's thinking

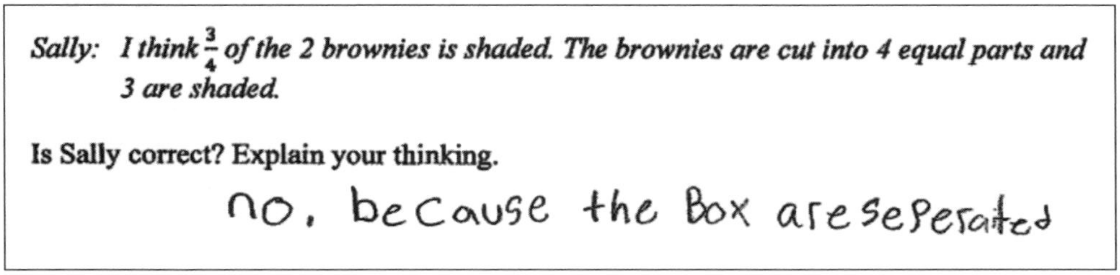

Sally: *I think $\frac{3}{4}$ of the 2 brownies is shaded. The brownies are cut into 4 equal parts and 3 are shaded.*

Is Sally correct? Explain your thinking.

no, because the Box are seperated

Fig. 2.4. Andy's response to Sally's thinking

Sally: *I think $\frac{3}{4}$ of the 2 brownies is shaded. The brownies are cut into 4 equal parts and 3 are shaded.*

Is Sally correct? Explain your thinking.

Yes, if you put them together it's $\frac{3}{4}$ like this.

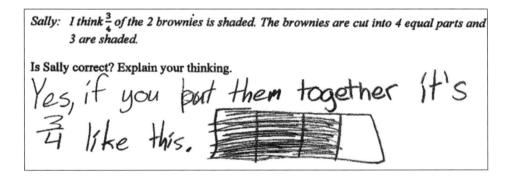

Fig. 2.5. David's response to Sally's thinking

Rafael appears to have a singular view of the unit—that is, a unit may involve only one item—and he seems to demonstrate difficulty in considering units other than those that are typically used in mathematics textbooks or in real-world contexts. He notes that the brownies are cut into "two wholes" and appears to ignore the language used by Sally that refers to the two brownies as the whole when she states that the shaded region is "$\frac{3}{4}$ of the 2 brownies."

It is worth noting that in our sample of fourth-, fifth-, and sixth-grade students, 35% correctly responded that Sally's response was right. However, some fourth- and fifth-grade students who responded correctly might have been applying low-level thinking and simply counting all the pieces to determine the denominator and counting the shaded pieces to determine the numerator—a process that would have resulted in their responding correctly. As Rafael's teacher, you could gain further insight into Rafael's thinking and understanding of the unit by asking him to draw diagrams of (a) $\frac{3}{4}$ of a brownie, (b) $\frac{3}{4}$ of 2 brownies, and (c) $\frac{3}{6}$ of 3 brownies. His responses could allow you to see whether he attended to the unit in each of these situations.

In addition to posing questions to further assess student thinking, you could draw on out-of-school situations to pose questions that would challenge and extend the thinking of students like Rafael. For example, consider Reese's Peanut Butter Cups, which are sold with 2 peanut butter cups to a package. If one person said, "I ate $\frac{1}{2}$ of a peanut butter cup," and another person said, "I ate $\frac{1}{4}$ of a package," who ate more? If one person said, "I ate $\frac{1}{2}$ of a peanut butter cup," and his friend said, "I ate $\frac{1}{2}$ of a package," who ate more? Having students draw diagrams to represent these amounts and discussing the similarities between the two diagrams could provide further insight. It is important to emphasize the unit regularly and stress that changing it requires adjusting the fractional amount that is identified.

Figure 2.4 shows the work of Andy, whose explanation of why Sally's response is incorrect is different from that offered by Rafael. He says that Sally is wrong because "the Box are separated." In our work, we found that many students responded in a manner similar to Andy, indicating the notion that the unit cannot be separated in a given diagram.

Note that Andy's response contrasts with David's suggestion, shown in figure 2.5, that putting the shaded parts together would allow saying that $\frac{3}{4}$ of the two brownies is shaded. In many mathematics textbooks, the unit is not explicitly stated in the text, and either the entire diagram or the portion of the diagram that is "connected" serves as the assumed unit.

Furthermore, units are sometimes unclear in the assignment portions of lessons, as illustrated in figure 2.6. The two tasks pictured in the figure make no mention of the unit, which means that students must make assumptions about it.

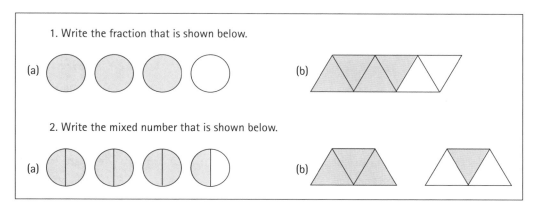

Fig. 2.6. Tasks that involve ambiguous units

By not specifying the unit in question, the items in the two tasks cause unnecessary confusion and reinforce the notion that one should "just know" what the unit is, on the basis of the spacing of the figures or other irrelevant information. From your experiences as a student and a teacher, you would probably expect that the answers to the problems in figure 2.6 are $\frac{3}{4}$, $\frac{4}{6}$, $3\frac{1}{2}$, and $1\frac{1}{3}$, respectively.

We encourage you to identify other correct answers that use different units from these expected answers. The use of ambiguous units and insufficient attention to units are problematic and can inhibit student understanding of multiplication and division of fractions. To encourage students to think further about the units in such tasks as these, you might initiate a discussion about the possible fraction names when various parts are "moved" in a diagram, as shown in figure 2.7.

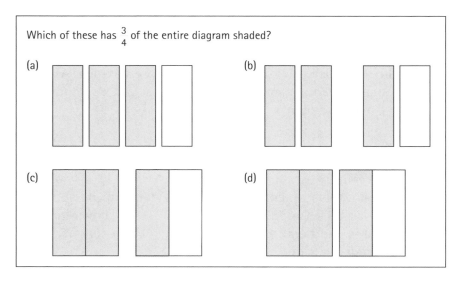

Fig. 2.7. A task that encourages students to consider the impact of location on the fraction name

In figure 2.7, each diagram in parts (a)–(d) could be denoted as $\frac{3}{4}$ of the entire diagram, but students who focus on the location of the parts might view only certain diagrams as $\frac{3}{4}$ of the whole diagram. Such a task can be used to emphasize that the physical location of the parts does not matter, since the task asks about the entire diagram, and the critical idea is how to view and refer to the unit in the situation. Students might continue to focus on the location and arrangement, considering, for example, whether the shaded regions are adjacent to determine the whole or unit. But continued discussions of these types of tasks can help students recognize that the arrangement of objects does not affect the unit.

How could you improve the wording of items that involve ambiguous units? Some textbook problems simply state, "Write the fraction that is represented in the picture." Other textbooks make reference to the diagram as the unit in statements such as, "Write the fraction of each figure or group." This is a step in the right direction. It might also be helpful during instruction to identify the unit explicitly, or to have students identify the unit and name the fraction. For example, you could say to your students, "Circle the part of the diagram that represents the unit, and then write a fraction or mixed number that represents the shaded portion." You could also identify different units for the same representation and ask students to identify the corresponding fraction for each unit.

Figures 2.8, 2.9, 2.10, and 2.11 show, respectively, the responses of four students, Maria (grade 5), Victoria (grade 5), Samantha (grade 6), and Serena (grade 6), to the answers of the fictitious students Marcus (1$\frac{1}{2}$ brownies are shaded) and Demetrius ($\frac{3}{2}$ of a brownie is shaded), shown in figure 2.1. Consider the questions in Reflect 2.3. As you examine the four students' evaluations of Marcus's and Demetrius's work, you may notice additional understandings and misunderstandings in the students' thinking about the relationships of the unit and the representation. As you read these students' responses, think about the strategies that you would use to move students forward from their current thinking.

Reflect 2.3

What mathematical understandings and misunderstandings do you find in the responses of Maria, Victoria, Samantha, and Serena, shown in figures 2.8–2.11, to Marcus's and Demetrius's thinking about the part of the brownies that is shaded in figure 2.1?

Reflect 2.3—Continued

As you look at these four students' responses, what similarities and differences do you see?

What specific strategies or questions would you use to move these students forward in their thinking?

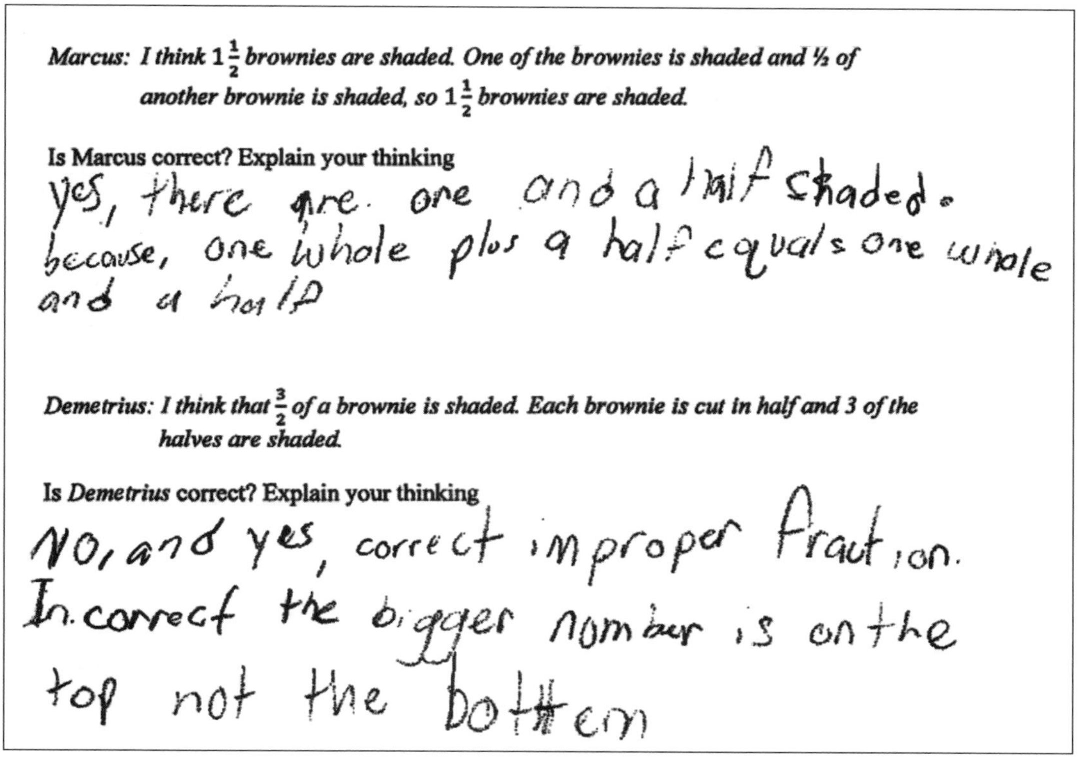

Marcus: I think $1\frac{1}{2}$ *brownies are shaded. One of the brownies is shaded and ½ of another brownie is shaded, so* $1\frac{1}{2}$ *brownies are shaded.*

Is Marcus correct? Explain your thinking

yes, there are one and a half shaded. because, one whole plus a half equals one whole and a half

Demetrius: I think that $\frac{3}{2}$ *of a brownie is shaded. Each brownie is cut in half and 3 of the halves are shaded.*

Is Demetrius correct? Explain your thinking

NO, and yes, correct improper fraction. In.correct the bigger number is on the top not the bottem

Fig. 2.8. Maria's responses to Marcus's and Demetrius's thinking

Maria appears to recognize that Marcus and Demetrius provided appropriate responses to this item, although she gives an answer of "yes and no" to the question of whether Demetrius is correct. She notes that Demetrius is incorrect because he does not provide a response in the correct form, since she apparently views an improper fraction as incorrect. Maria focuses on the form rather than the meaning in this situation, unfortunately. Demetrius provided a correct response, and the form of the response does not matter. It is essential that students recognize and apply the

> **Marcus:** *I think* $1\frac{1}{2}$ *brownies are shaded. One of the brownies is shaded and ½ of another brownie is shaded, so* $1\frac{1}{2}$ *brownies are shaded.*
>
> Yes, Marcus is correct. 1½ brownies are shaded. I whole-and 1 half. He is correct even though he could've answered differently.
>
> **Demetrius:** *I think that* $\frac{3}{2}$ *of a brownie is shaded. Each brownie is cut in half and 3 of the halves are shaded.*
>
> **Is Demetrius correct? Explain your thinking**
>
> Yes, he is correct, also. There are 2 halves in one whole which is shown. There is still a left over of one half which totals to 3 halves. I is a improper fraction

Fig. 2.9. Victoria's responses to Marcus's and Demetrius's thinking

meaning of $\frac{3}{2}$ as 3 parts, each of which is $\frac{1}{2}$. In other words, they must recognize that a piece that is $\frac{1}{2}$ of the unit has been *iterated*, or repeated, 3 times.

It is important to emphasize in the classroom that either "$1\frac{1}{2}$ brownies" or "$\frac{3}{2}$ of a brownie" is an acceptable response to this task, since these responses refer to equivalent amounts. Victoria's response demonstrates that she recognizes that each shaded part is $\frac{1}{2}$ of a brownie, and 3 half-brownies are shaded. In addition to showing a level of comfort in working with both mixed numbers and improper fractions, Victoria demonstrates an understanding of *iteration*—in this case, the process of starting with a part that is equal to $\frac{1}{2}$ and repeating it twice to make a whole and then once more to make $\frac{3}{2}$.

Marcus: I think $1\frac{1}{2}$ brownies are shaded. One of the brownies is shaded and ½ of another brownie is shaded, so $1\frac{1}{2}$ brownies are shaded.

Is Marcus correct? Explain your thinking

Marcus is correct because 1 whole is shaded and $\frac{1}{2}$ is shaded

Demetrius: I think that $\frac{3}{2}$ of a brownie is shaded. Each brownie is cut in half and 3 of the halves are shaded.

Is *Demetrius* correct? Explain your thinking

Demitrius is incorrect because 1 whole is shaded and there is $\frac{1}{2}$ half that is shaded

Fig. 2.10. Samantha's responses to Marcus's and Demetrius's thinking

Marcus: I think $1\frac{1}{2}$ brownies are shaded. One of the brownies is shaded and ½ of another brownie is shaded, so $1\frac{1}{2}$ brownies are shaded.

Is Marcus correct? Explain your thinking

yes, because one box is shaded all in and the other one is shaded so it is $\frac{1}{2}$.

Demetrius: I think that $\frac{3}{2}$ of a brownie is shaded. Each brownie is cut in half and 3 of the halves are shaded.

Is *Demetrius* correct? Explain your thinking

no, beause their are 4 picces

Fig. 2.11. Serena's responses to Marcus's and Demetrius's thinking

Samantha and Serena appear to have difficulties that would be more difficult to overcome than Maria's. Samantha appears to look at the diagram in only one way, as representing $1\frac{1}{2}$, and not as showing $\frac{3}{2}$, as well. Serena also writes that $\frac{3}{2}$ is incorrect, pointing to the fact that there are 4 parts visible. Both students might have difficulty thinking of the unit in relation to the denominator. They might count the shaded parts and the total parts or refer to familiar diagrams. They might not have coordinated the meaning of the fraction with the unit. In this case, these students would be likely to have difficulty with tasks requiring them to determine the whole or unit when given part of the whole. One such task is shown in figure 2.12.

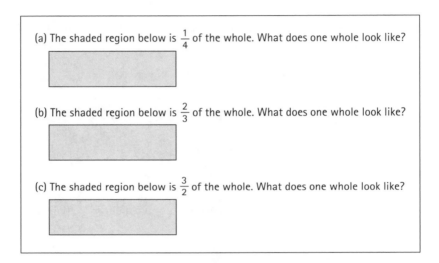

Fig. 2.12. A task that requires creating the unit, given a particular fraction of the unit

To be successful with tasks such as these, students must understand the meaning of the numerator and the denominator of a fraction and apply these ideas appropriately. Unfortunately, textbooks and online resources too often refer to the denominator as the "bottom number" and the numerator as the "top number," rather than emphasizing the meanings of these terms. Van de Walle (2007, p. 300) describes this shortcoming:

> The bottom number is said to tell "how many parts it takes to make a whole." This may be correct but can be misleading. For example, a $\frac{1}{6}$ piece is often cut from a cake without making any slices in the remaining $\frac{5}{6}$ of the cake. That the cake is only in two pieces does not change the fact that the piece taken is $\frac{1}{6}$.

Van de Walle explains that the numerator "counts" and the denominator is "the name for what is being counted."

Thus, to be successful with the task shown in figure 2.12, students need to be able to iterate and partition shapes. For example, in part (a), students need to iterate, or repeatedly copy, the given $\frac{1}{4}$ three times to create a total of 4 parts to form the whole. In part (b), they need to partition the region of $\frac{2}{3}$ into 2 parts, each of which they understand as representing $\frac{1}{3}$, and then they must iterate the $\frac{1}{3}$ piece to give them three $\frac{1}{3}$ pieces to make a whole. Siebert and Gaskin (2006) and Battista (2012) offer further discussion of iterating and partitioning fractions to help students develop a deeper understanding of the meaning of fractions.

It is also important to note that an improper fraction such as $\frac{3}{2}$ often causes difficulties for students because the meaning of fractions like these can be challenging to connect with the quantities that these fractions represent. One strategy for addressing this difficulty is to give students opportunities to discuss and draw images of "half of an orange," "two halves of an orange," and "three halves of an orange," for example, or "half of a foot of ribbon," "two halves of a foot of ribbon," and "three halves of a foot of ribbon." Students—and adults—often make incorrect assumptions that the reference to the unit in phrases such as " $\frac{3}{2}$ of an orange" means that only one orange can be considered, since the statement indicates "*an* orange."

However, it is important for students and adults alike to recognize that the phrase "of an orange" in " $\frac{3}{2}$ of an orange" names the whole, or unit, for the fraction $\frac{3}{2}$. Therefore, " $\frac{3}{2}$ of an orange" would consist of three parts, with each part equal to $\frac{1}{2}$ of an orange. Some students think that fractions must always represent quantities that are less than one whole or unit, and many of these students fail to establish a meaning for fractional quantities (other than mixed numbers) that are greater than 1.

Summarizing Pedagogical Content Knowledge to Support Essential Understanding 2*a*

Teaching the mathematical ideas in this chapter requires specialized knowledge related to the four components presented in the Introduction: learners, curriculum, instructional strategies, and assessment. The four sections that follow summarize some examples of these specialized knowledge bases in relation to Essential Understanding 2*a*. Although we separate them to highlight their importance, we also recognize that they are connected and support one another.

Knowledge of learners

Battista (2012, p. 2) emphasizes students' need to understand the processes of partitioning and iterating—interrelated processes that enable students to apply and develop their concept of a unit for a fraction:

> Deeper understanding of the role of partitioning in fractions comes from understanding the complementary process of iteration. Partitioning starts with the whole and divides it into equal parts. Iteration starts with a part and repeats it to make the whole. Students take a major step toward substantive understanding of fractions when they understand the relationship between the processes of iteration and partitioning.

The examples of students' work included in this chapter have illustrated the value of these processes. As we consider what we know about Samantha and Serena, the two sixth-grade students whose thinking is shown in figures 2.10 and 2.11, we recognize that they need support in developing essential understandings related to unit and improper fractions. Designing instruction, assessment, and curricular tasks that enhance student understanding relies on our specific knowledge of learners' mathematical understandings and misconceptions. Thus, providing tasks and posing questions about the process of iteration and its relationship to fractions will help students like Samantha and Serena develop important foundational understandings.

Knowledge of curriculum

Selection of important ideas to be addressed in mathematical tasks is critical to robust student learning. The tasks in this chapter have focused on the following:

- The interpretation of a given unit and its symbolic fraction representation

- The impact of location and separation on the fraction name

- The construction of the unit, given a fractional amount

In addition to carefully selecting tasks that will not perpetuate the misconceptions that lie behind students' faulty concept images, you may need to enhance tasks found in curricular materials. As discussed, you may need to highlight the unit, as illustrated in figure 2.1, or ask children to identify the unit prior to solving a problem. Furthermore, textbook problems often have limitations, as Kabiri and Smith (2003, p. 187) explain:

> Textbook problems do not always lend themselves to multiple solutions, or solution strategies. However, many problems can be made more open-ended and accessible to a wide variety of student ability levels with minimum effort.

As a teacher, you must consider your curricular goals and the students that you teach as you enhance mathematical tasks in your curricular materials. For example, do your materials explicitly address the significance of units? Are tasks designed in ways that facilitate analysis and debate of student misconceptions? Do the tasks encourage multiple solutions or solution strategies? Is the language misleading or confusing? Ideas for enhancing curricular materials are available in Chval and Davis (2008), Chval and colleagues (2009), Chval and Chávez (2012), and Kabiri and Smith (2003).

Knowledge of instructional strategies

Many instructional strategies are possible for teaching fractions. Chapter 1 introduced the strategy of having students evaluate the work of a fictitious student who has a misconception (see fig. 1.10). Chapter 2 has introduced a variation on that idea—the idea of having students examine the work of fictitious students who display correct thinking. Figure 2.1 presents three fictitious students' conjectures involving the important idea of unit. Although the fictitious students display correct thinking, having your own students examine their work on the task can help you bring to the surface misconceptions that they may hold in relation to unit.

Another strategy that you can use to help your students develop their understanding of the unit is to identify the unit explicitly in the problems that you use. The emphasis that this strategy gives to the unit is appropriate, since the unit, or whole, is the foundation for the fraction concepts that students develop.

Knowledge of assessment

As noted in the Introduction, Wiliam (2007) emphasizes the importance of selecting tasks that provide opportunities for students to demonstrate their thinking and for teachers to make instructional decisions. Assessment tasks, like instructional tasks, should routinely call on students to create and interpret diagrams. As you consider the assessment in your school or district, ask yourself whether the tasks allow you the opportunity to understand the way—or the multiple ways—in which your students think about units. Do the assessment tasks focus on the big ideas related to the concept of unit? When are students interpreting fraction diagrams? When are they creating their own fraction diagrams? How do the responses to the tasks help you craft your next instructional moves?

Conclusion

Identifying the unit and recognizing its connection with a fractional part are essential to understanding the meaning of fractions. Students who are developing this understanding need special support and guidance. Offering them tasks with ambiguous units and giving insufficient attention to units are problematic, potentially inhibiting their understanding and impeding their future work with computation of fractions. It is helpful during instruction to identify the unit explicitly or to have students identify it and name the fraction. It is also helpful to promote students' understanding of the relationship between the processes of iterating and partitioning. Developing these ideas requires a careful selection of tasks and effective questions, but teachers who make this effort will enable their students to lay a foundation for developing an essential understanding of different interpretations of fractions—the topic of the next chapter.

Chapter 3
Interpretations of Fractions

Essential Understanding 2*b*
One interpretation of a rational number is as a part–whole relationship.

Essential Understanding 2*c*
One interpretation of a rational number is as a measure.

The effective use of models, tools, and representations is a key component of all successful mathematics teaching (Clarke and Clarke 2003). Clearly, helping your students develop a deep understanding of fractions requires giving them opportunities to consider and work with various representations and interpretations. *Developing Essential Understanding of Rational Numbers for Teaching Mathematics in Grades 3–5* (Barnett-Clarke et al. 2010) presents interpreting a rational number as a relationship of a part to a whole and as a measure as Essential Understandings 2*b* and 2*c*. Ensuring that students become accustomed to interpreting fractions in these ways through the use of models is the focus of this chapter.

Working toward Essential Understandings 2*b* and 2*c*

Research cautions us about the consequences when students lack experience with different representations and interpretations of fractions. Multiple studies demonstrate the difficulties that arise when students do not develop a deep understanding of the underlying mathematical ideas that can be viewed through various models and representations.

For example, consider the use of models that are "pre-labeled" with fraction symbols that have meaning in another context but not in the one at hand. Figure 3.1 shows a task presented with such pre-labeled tools. Many students would be likely to respond that the upper rod is $\frac{1}{2}$ of the length of the lower rod, since the upper rod is labeled as $\frac{1}{2}$. However, the length of the upper rod is in fact $\frac{2}{3}$ of the length of the lower rod.

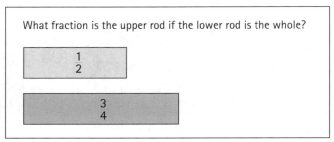

Fig. 3.1. Working with "pre-labeled" fraction tools in a task comparing two lengths

Lesh, Landau, and Hamilton (1983) found that students who obtained incorrect answers for their written calculations could often determine the correct answer through the use of an appropriate model. However, even after these students had obtained the correct answers by using the model, when they were confronted with their incorrect written symbolic work, about half of them thought it was correct. This discrepancy in their views of their results from working in two different contexts reveals that students often have difficulty making connections between formal, abstract mathematical symbols and less formal, more concrete models of them.

In an encouraging study, Wearne and Hiebert (1988) found that fourth graders whose instruction focused on connecting decimal fractions with physical representations were more successful in dealing with problems that they had not seen before—such as ordering decimals and converting between decimal and common fraction forms—than students who had not received such instruction. Thus, if instruction focuses on developing a deeper understanding of rational numbers through the use of various models, students are more likely to apply their understanding in new situations. However, if instruction fails to help students connect symbolic representations with models of the corresponding fractions, students may have difficulty overcoming their misconceptions about the meaning of fractions.

To help students develop a deeper understanding of fractions, it is valuable to use both pictorial and symbolic representations in the classroom. Even though these representations can support appropriate understandings, they can also lead to misunderstandings. Being aware of these can be helpful when you are introducing your students to concepts related to fractions.

Researchers have identified three fundamental types of models that can help students develop their understanding of the meaning of fractions and operations with fractions:

1. Linear models, typically modeled with tools such as number lines or Cuisenaire rods

2. Area models, typically modeled with tools such as rectangular fraction bars, fraction circles, or pattern blocks

3. Discrete models, including any model viewed as a group of objects, such as the children in the classroom or the cars in the parking lot, with each object in the group typically represented by some type of counter

Watanabe (2002) and Van de Walle (2007) provide further discussion of these models. It is important to recognize that neither the arrangement of the pictorial representation nor the pictorial representation itself determines whether a model is a linear, area, or discrete model. Instead, the type of model is indicated by the context or point of view of the individual using it. For example, a student may view a fraction strip as an area model or a linear model.

The previous chapter noted that many textbooks and authors do not clarify the unit in tasks for students. Likewise, many do not clarify the type of model that they are providing, leaving it open to various interpretations. As Watanabe (2002) states, "Nothing inherent in pattern blocks prevents children from modeling fractions 'discretely' by using each piece as the unit of counting" (p. 458). Thus, though your intention might be for your students to view a figure as an area model, they might view it discretely. Wu (2012) also notes that we need to clarify the attribute when we discuss fractions. For example, sometimes we refer to "$\frac{1}{2}$ of a circle," when we really mean "$\frac{1}{2}$ of the area of a circle." We make similar statements in our everyday language, for example referring to "$\frac{1}{2}$ of a dollar" or "$\frac{3}{4}$ of the water." Our implied meaning is "$\frac{1}{2}$ of the amount of money in a dollar" or "$\frac{3}{4}$ of the volume of the water," but which attribute we are referring to may not be clear to our students.

Figure 3.2 presents the Fraction Hexagon task, which calls for an evaluation of the work of two fictitious students with an area model. As the fictitious students' responses show, the model lends itself to multiple interpretations. Reflect 3.1 poses questions to guide your examination of the task. (Note that in commercially produced pattern blocks, the triangle is customarily green, the rhombus is blue, and the trapezoid is red. The areas of the triangle, rhombus, and trapezoid are, respectively, one-sixth, one-third, and one-half of the area of the hexagon.)

Reflect 3.1

What aspects of student understanding about fractions are being assessed in the Fraction Hexagon task, shown in figure 3.2?

How would grade 3–5 students respond to this task?

Which of the two fictitious students, Mary or Michelle, provided a correct fraction name and referred to the correct corresponding unit?

Why do you think these responses were chosen for the task?

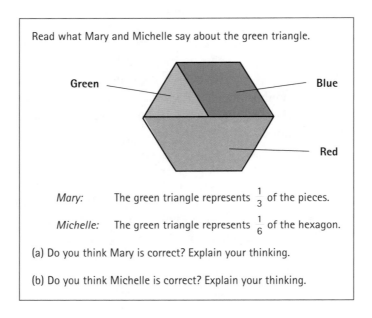

Read what Mary and Michelle say about the green triangle.

Green — Blue

— Red

Mary: The green triangle represents $\frac{1}{3}$ of the pieces.

Michelle: The green triangle represents $\frac{1}{6}$ of the hexagon.

(a) Do you think Mary is correct? Explain your thinking.

(b) Do you think Michelle is correct? Explain your thinking.

Fig. 3.2 Fraction Hexagon task

The Fraction Hexagon task can provide insight into how students view this representation. Mary views the diagram discretely and refers to the total number of pieces as the whole. Though many would say that Mary is incorrect, the diagram does not clarify that an area model is the intended representation. Thus, Mary's discrete view of the diagram, confirmed by her use of "of the pieces" in her specification of her reference unit, is a plausible view. Michelle provides what many would view as the "correct" fraction for this diagram. Michelle considers the area of the pieces in determining that the green triangle represents $\frac{1}{6}$ of the hexagon.

Students should be able to view this model as a discrete model or as an area model, depending on the situation.

You may now be thinking, "How could students ever know whether a model is to be viewed as an area model or as a discrete model?" The context of the situation plays a critical role in clarifying the intended view of the model. For example, if a task presents a context in which the students are given a two-dimensional diagram and asked to find "the fraction of a cookie," the situation suggests that they are to view the model as an area model. However, such contexts may not prevent students from viewing representations discretely. Consequently, it is important that you emphasize the need to examine the context of the situation to determine whether a model can be viewed discretely. Unfortunately, many textbooks do not clarify the intended type of model but instead induce students to view a model in the way that the textbook's authors prefer through the answer key.

Figure 3.3 provides an overview of how students entering grades 3–5 responded to the Fraction Hexagon task shown in figure 3.2. It is interesting to note the development in students' responses between third and fifth grade.

	Mary is correct	Michelle is correct
Third grade ($n = 25$)	60% $\left(\frac{15}{25}\right)$	20% $\left(\frac{5}{25}\right)$
Fourth grade ($n = 30$)	47% $\left(\frac{14}{30}\right)$	53% $\left(\frac{16}{30}\right)$
Fifth grade ($n = 27$)	44% $\left(\frac{12}{27}\right)$	67% $\left(\frac{18}{27}\right)$

Fig. 3.3. Students' responses to Mary's and Michelle's responses in the Fraction Hexagon task

Specific examples of student work can provide further insight into the mathematical thinking that can emerge in response to the Fraction Hexagon task. The questions in Reflect 3.2 direct your attention to figures 3.4, 3.5, and 3.6, which show the work of Gerald (entering grade 5), Diana (entering grade 4), and Martha (also entering grade 4), respectively, on the task. Consider their responses to the work of Mary and Michelle, shown in figure 3.2.

Reflect 3.2

What mathematical understandings and misunderstandings do you find in the thinking of Gerald, Diana, and Martha, whose work on the Fraction Hexagon task (in fig. 3.2) is shown in figures 3.4, 3.5, and 3.6, respectively?

What specific strategies would you use to move these students forward?

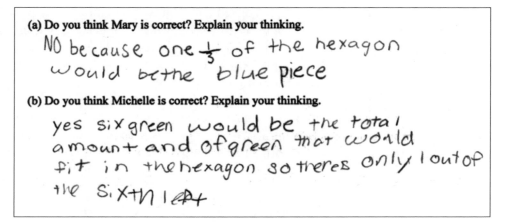

(a) Do you think Mary is correct? Explain your thinking.

NO because one ⅓ of the hexagon would be the blue piece

(b) Do you think Michelle is correct? Explain your thinking.

yes six green would be the total amount and of green that would fit in the hexagon so theres only 1 out of the sixth left

Fig. 3.4. Gerald's response to the thinking of Mary and Michelle

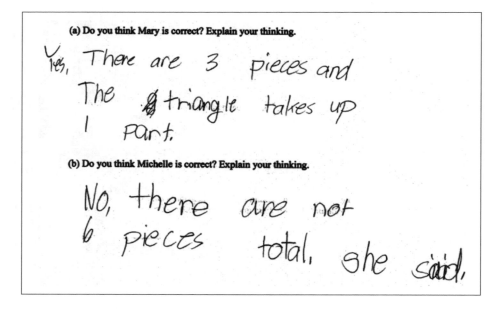

(a) Do you think Mary is correct? Explain your thinking.

Yes, There are 3 pieces and The ⅓ triangle takes up 1 Part.

(b) Do you think Michelle is correct? Explain your thinking.

No, there are not 6 pieces total, she said.

Fig. 3.5. Diana's response to the thinking of Mary and Michelle

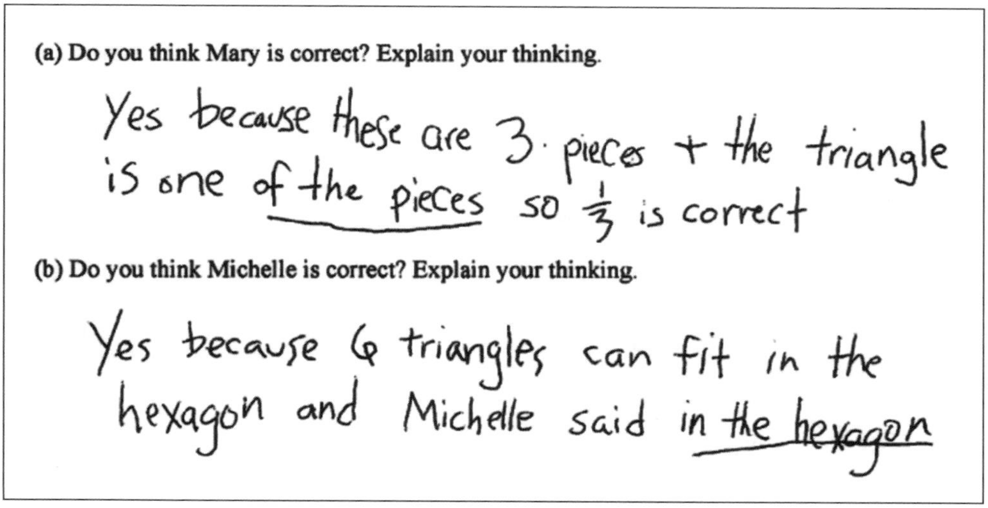

(a) Do you think Mary is correct? Explain your thinking.

Yes because these are 3. pieces + the triangle
is one of the pieces so $\frac{1}{3}$ is correct

(b) Do you think Michelle is correct? Explain your thinking.

Yes because 6 triangles can fit in the
hexagon and Michelle said in the hexagon

Fig. 3.6. Martha's response to the thinking of Mary and Michelle

Gerald represents the students who believed that Mary was incorrect and Michelle was correct. In fact, nearly half of the students in grades 4 and 5 (13 out of 30 and 13 out of 27, respectively) agreed with him. He interpreted this task solely as an area model. He did not consider interpreting the diagram as a discrete model but instead suggested that the "blue piece"—the rhombus—would be $\frac{1}{3}$ of the hexagon. Jackson, a student entering grade 3 whose work is not shown, also believed that Mary's answer was incorrect. He offered the following explanation for this conclusion: "No, [she is not correct] because fractions have to be the same size and they are not."

By contrast, Diana, a student entering grade 4, was able to interpret the diagram as a discrete model, but she had difficulty expanding her view to incorporate the idea of it as an area model. The majority of the grade 3 students (13 of 25) in the study shared this limited view, which was less common among students in grade 4 (10 out of 30) and grade 5 (7 out of 27). Diana interpreted Michelle's statement as incorrect because she viewed the denominator strictly as the number of pieces needed—not as an area.

Martha, also a student entering grade 4, showed greater flexibility in using different models and units to evaluate Mary's and Michelle's work on the task. Martha was willing to say that the answers provided by Mary and Michelle are both correct, depending on students' interpretations of the models that they used. Eight students in our study (3 out of 30 fourth graders and 5 out of 27 fifth graders) gave similar responses.

Another characteristic of discrete models that distinguishes them from area or length models is that the objects composing the unit need not be the same size. For example, a discrete set of objects might have a unit made up of a group of people or a group of coins—say, pennies, nickels, and dimes. A discrete set might consist of all the people in a room, $\frac{1}{3}$ of whom are wearing sandals, or a collection of coins in a bowl, $\frac{1}{4}$ of which are pennies. Neither the people in the room nor the coins in the bowl would necessarily be the same size, since this is not a requirement for a discrete model. However, the assumption with area and length models is that these models are partitioned into parts of the same size, although with area models, the parts need not be the same shape.

What can we say about students' difficulties with each of these models? Watanabe (2002) describes a task presented to fifth-grade students. It showed them three cookies, represented by shaded pieces of a square (see fig. 3.7). The students were first shown that two congruent shaded pieces composed the square. Then the students were asked to decide which "cookie" they would select if they were really hungry. Although each shaded portion represented the same fraction ($\frac{1}{2}$) of the square in each of the three diagrams, most of the students (14 of 16 students) chose one particular diagram over the others, demonstrating that students' lack of understanding of area can interfere with their understanding of fractions. Accordingly, Watanabe cautions teachers on the use of area models in the elementary grades, when students are still developing an understanding of area.

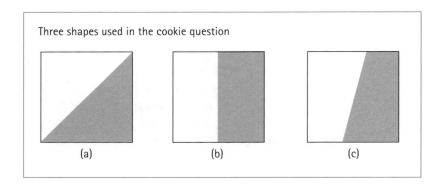

Fig. 3.7. Cookie shapes (shaded portions of squares) from Watanabe (2002)

Research has shown that length models present their own difficulties for students who are learning about fractions. With respect to the use of number lines, Shaughnessy (2011) found that fifth-grade students demonstrated difficulty in placing various decimals and fractions on number lines. She identified four types of errors that these students made:

1. Using unconventional notation. Students demonstrated (as Chapter 5 notes) a misunderstanding of formal notation for fractions and decimals.

2. Redefining the unit. Students ignored the unit marked from 0 to 1 and inadvertently renamed the unit (see fig. 3.8).

3. Applying a "two-count strategy" focusing on discrete tick marks rather than distances. Students counted all the tick marks for the units (rather than counting the distance between the tick marks) and counted the number of tick marks to the reference point (see fig. 3.9).

4. Applying a "one-count strategy" focusing on discrete tick marks rather than distances. Students simply used a single count to determine, incorrectly, the fraction at a particular point (see fig. 3.10).

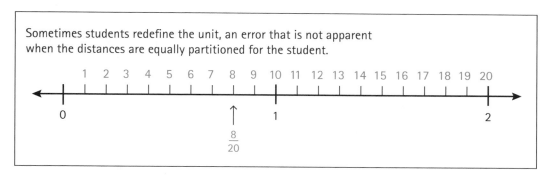

Fig. 3.8. Misplacing a fraction as a result of renaming the unit (Shaughnessy 2011, p. 432)

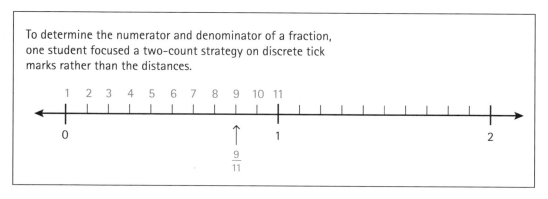

Fig. 3.9. Misplacing a fraction as a result of applying a two-count strategy focusing on tick marks (Shaughnessy 2011, p. 433)

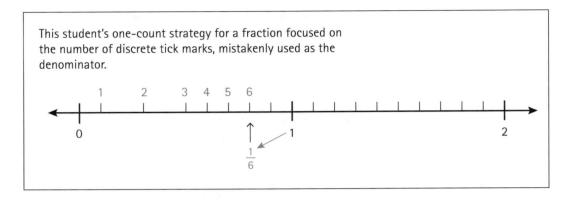

Fig. 3.10. Misplacing a fraction as a result of applying a one-count strategy focusing on discrete tick marks rather than distances (Shaughnessy 2011, p. 433)

Mack (1990) and Watanabe (2002) also provide helpful considerations related to the use of formal fraction symbols in notation like "$\frac{3}{4}$" and "$\frac{6}{9}$." Mack describes the rich prior knowledge about partitioning that students bring to the classroom from their out-of-school experiences with fractions. However, when fraction symbols are introduced in school, students frequently draw on their prior knowledge of whole numbers—knowledge that interferes with their understanding of fraction operations.

For example, when students see the fraction $\frac{3}{4}$, they see a "3" and a "4," and these symbols naturally draw on their previous understanding of the meaning of the whole numbers 3 and 4. Mack notes that such interference appears to be inevitable but can be overcome through a focus on the meaning of fraction notation. An understanding of the meaning of the notation is critical before the students begin working with operations on fractions.

To aid students in developing an understanding of the meaning of the numerator as a count of the parts making up the whole and the denominator as the name for the part of the whole that is under consideration, Gunderson and Gunderson (1957) suggest that teachers initially refer to $\frac{3}{4}$ as "3-fourths" to emphasize that $\frac{3}{4}$ means 3 parts that are called "fourths." In addition, Siebert and Gaskin (2006) recommend that teachers avoid language such as "three over four" and "three out of four" because this language can perpetuate students' views of numerators and denominators as separate values.

Summarizing Pedagogical Content Knowledge to Support Essential Understandings 2*b* and 2*c*

Teaching the mathematical ideas in this chapter requires specialized knowledge related to the four components presented in the Introduction: learners, curriculum, instructional strategies, and assessment. The four sections that follow summarize some examples of these specialized knowledge bases in relation to Essential Understandings 2*b* and 2*c*. Although we separate them to highlight their importance, we also recognize that they are connected and support one another.

Knowledge of learners

As demonstrated by Gerald's, Diana's, and Martha's evaluations of Mary's and Michelle's work in figure 3.2 on the Fraction Hexagon task, students may view a fraction model as a different type (area, linear, or discrete) from the type that the teacher (or the textbook) intended. For example, you may draw a part-whole representation, expecting your students to consider it as an area model, only to discover that they view it—and work with it—as a discrete model.

Therefore, it is critical for you to determine how your students view different fraction models and representations that are used in the classroom. In addition, you need to determine whether students have sufficient understanding of the underlying mathematical concepts in different fraction models. For example, Watanabe (2002) cautions against using area models when elementary students are still developing an understanding of area. Battista (2012, p. 74) outlines important ideas about student understanding of number lines:

> First, students must understand the segment between 0 and 1 on the number line should be partitioned into equal [length] sections. So students must understand that counting hash marks or unequal segments is incorrect.

Chapter 4 focuses on different challenges that emerge in using different models and representations to help children develop their understanding of fractions.

Knowledge of curriculum

It is not unusual for textbooks to emphasize area models and fraction symbols rather than other models, such as discrete and linear models (Empson 2002). Which models are emphasized in the curricular materials that you use? How do the lessons in these curricular materials help students build their understanding of the meaning of these different models? Do students have opportunities to create their own representations or discuss the meaning of the different components of a representation? Carefully selecting, enhancing, and designing appropriate mathematical tasks requires attending not only to students' understandings and misconceptions—along with the contexts, units, and language in the tasks—but also to the mathematical models and representations that the tasks use and how students are likely to perceive them.

Knowledge of instructional strategies

Teachers can draw on many and varied strategies in helping their students interpret fractions and work with models and representations of them. We highlight just two examples here. The first is to give students opportunities to create their own models and representations—a strategy that Empson (2002, p. 35) suggests:

> Allowing children to choose their own tools and make their own representations ... can foster an interesting diversity of thinking, which can contribute to richer understanding of the mathematics of fractions.

The second strategy that we highlight is to break down and focus on the components of all models and representations in using them with students. As you facilitate discussions about your students' representations, emphasize the meaning of their various components. As your students engage in a task, pose questions such as the following:

- What does this [*using a gesture to indicate a specific part*] represent?

- Where, or what, is the unit, or whole, in your representation?

- How does your representation differ from your partner's representation? How is your representation similar to your partner's?

Knowledge of assessment

Figures 3.8–3.10 demonstrate different incorrect strategies that many students use in attempting to identify fractions on a number line. Carefully designed assessment tasks can make these errors visible, giving teachers information that they can use to help students develop accurate understandings. For example, the student whose work is shown in figure 3.8 uses 20 as the denominator. If this student had been given a number line labeled only between 0 and 1, he would not have used 20 as the denominator, and the teacher might have lost an opportunity to identify this misconception.

Another potentially productive strategy is to include an atypical representation in an assessment. For example, a representation such as (c) in figure 3.7, which shows an unconventional division of a square, could help students develop essential understandings related to area and area models of fractions. Anticipating common errors or misconceptions can help you select or design appropriate assessment tasks.

Conclusion

To develop a deep understanding of fractions, students need to consider and work with various representations and interpretations. These include different types of models (area, linear, and discrete) as well as both pictorial and symbolic representations of fractions. Even though working with models and representations can support appropriate understandings, students may interpret them in different ways, possibly leading to misunderstandings. For example, your students may see a fraction model as a different type of model from the type that you—or your textbook—intended. Or they may misunderstand how to use a model, possibly exhibiting the confusion shown in the work with number lines in figures 3.8–3.10. By carefully selecting tasks and posing effective questions, you will help your students understand ways of interpreting fractions, laying the foundations for them to understand equivalence and comparison of fractions, the topics discussed in the next chapter.

into practice

Chapter 4
Equivalence and Comparison of Fractions

Essential Understanding 3*a*
Any rational number can be expressed as a fraction in an infinite number of ways.

This chapter highlights two important components of the idea that any rational number can be expressed as a fraction in an infinite number of ways, the concept presented as Essential Understanding 3*a* in *Developing Essential Understanding of Rational Numbers for Teaching Mathematics in Grades 3–5* (Barnett-Clarke et al. 2010). The two facets of this idea on which this chapter focuses are equivalent fractions and the comparison of fractions.

Working toward Essential Understanding 3*a*

Understanding equivalent fractions and comparing fractions pose unique new challenges to students in grades 3–5. They require understanding fundamental ideas related to the meaning of fractions through the use of various representations. Although concepts of equivalence and comparison of fractions involve similar understandings, the discussion that follows considers them separately, focusing first on tasks reinforcing understanding of equivalent fractions and then turning to tasks supporting comparisons of fractions.

Equivalent fractions

Essential Understanding 3*a* describes a characteristic of rational numbers that makes them unlike any other numbers that elementary students have encountered in mathematics: rational numbers include mathematical objects that do not look the same but represent equivalent amounts—for example, $\frac{3}{4}$ and $\frac{6}{8}$.

As noted by Barnett-Clarke and colleagues (2010), the words "equivalent" and "equal" mean essentially the same thing, yet students may give different meanings to them and to a third term—"same." These differences may stem from thinking of representations of physical quantities, such as $\frac{3}{4}$ and $\frac{6}{8}$ of the same thing, as "equivalent" amounts or values, but reading the equation $\frac{3}{4} = \frac{6}{8}$ aloud as "$\frac{3}{4}$ equals $\frac{6}{8}$." Moreover, students may hear the statement, "Three-fourths is the same as $\frac{6}{8}$," but they can plainly see that "$\frac{3}{4}$" does not look the same as "$\frac{6}{8}$." It is challenging to use language carefully while helping students develop an understanding of the associated meanings of the terms "equal," "equivalent," and "same."

Depending on the grade level, some students may have experiences with the word "equivalent," and some of these students may understand what it means, but others may be unfamiliar with the term. For example, some students may have first learned about equivalence when they were introduced to equivalent amounts of money—such as 5 dimes and 2 quarters. Students may have thought of equivalent amounts as different names for the same amount as they responded to questions such as, "What are different names for $\frac{1}{2}$?" (Van de Walle 2007).

As a result of these differences in students' experiences, it is important for you to assess your students' current thinking about equivalence before using it in a fraction context. Bassarear (1997) argues, "Almost all of the essential fraction ideas surface in the concept of equivalent fractions, and this concept permeates many of the fraction problems that we encounter" (p. 222). Therefore, children who do not understand equivalence will struggle with essential understandings associated with rational numbers.

Often textbooks state that $\frac{3}{4}$ is equivalent to $\frac{6}{8}$ without further discussion of the underlying assumption that this is true if and only if $\frac{3}{4}$ and $\frac{6}{8}$ both refer to the same unit. For example, $\frac{3}{4}$ of \$100 is not the same value as $\frac{6}{8}$ of \$50, but $\frac{3}{4}$ of a candy bar is the same amount as $\frac{6}{8}$ of the same candy bar. The underlying assumption about referring to the same unit also applies to comparing any two fractions. When comparing two fractions, we assume that both refer to the same unit. This is an important point to emphasize when you and your students compare fractions and discuss fraction equivalence.

Another critical idea that students must develop for a deep understanding of fraction equivalence is that two fractions that do not look the same may in fact represent equivalent amounts. The Common Core State Standards for Mathematics (CCSSM) emphasize this point, stating that students in third grade should be able to "recognize and generate simple equivalent fractions, e.g., $\frac{1}{2} = \frac{2}{4}$, $\frac{4}{6} = \frac{2}{3}$ ... [and] explain why the fractions are equivalent, e.g., by using a visual fraction model" (National Governors Association Center for Best Practices and Council of Chief State School Officers [NGA Center and CCSSO] 2010, p. 24). Figure 4.1 shows the complete statement of this grade 3 standard in CCSSM.

Common Core State Standards for Mathematics, Grade 3

Develop understanding of fractions as numbers.

3. Explain equivalence of fractions in special cases, and compare fractions by reasoning about their size.

    a. Understand two fractions as equivalent (equal) if they are the same size, or the same point on a number line.

    b. Recognize and generate simple equivalent fractions, e.g., $\frac{1}{2} = \frac{2}{4}$, $\frac{4}{6} = \frac{2}{3}$. Explain why the fractions are equivalent, e.g., by using a visual fraction model.

    c. Express whole numbers as fractions, and recognize fractions that are equivalent to whole numbers. *Examples: Express 3 in the form* $3 = \frac{3}{1}$; *recognize that* $\frac{6}{1} = 6$; *locate* $\frac{4}{4}$ *and 1 at the same point of a number line diagram.*

    d. Compare two fractions with the same numerator or the same denominator by reasoning about their size. Recognize that comparisons are valid only when the two fractions refer to the same whole. Record the results of comparisons with the symbols >, =, or <, and justify the conclusions, e.g., by using a visual fraction model.

Fig. 4.1. Number and Operations—Fractions, CCSSM 3.NF.3
(NGA Center and CCSSO 2010, p. 24)

Kamii and Clark (1995) argue that even when students can generate equivalent fractions by working with physical models, they are not necessarily able to explain why these fractions are equivalent. In fact, many adults learned a rule for generating equivalent fractions when they were younger without understanding why this rule works.

Figures 4.2, 4.3, 4.4, and 4.5, show, respectively, the explanations offered by students A, B, C, and D of why $\frac{3}{4}$ is the same amount as $\frac{6}{8}$. Which explanations, if any, are valid, and which are invalid? Use the questions in Reflect 4.1 to help you gain further insight into students' understanding or misunderstanding of the underlying meaning of equivalent fractions.

Reflect 4.1

How would you characterize the mathematical understandings and misunderstandings that students A–D exhibit in figures 4.2–4.5 in explaining why $\frac{3}{4}$ is the same amount as $\frac{6}{8}$?

For each explanation that you think is invalid, address the misconceptions that invalidate it.

What specific strategies would you use to move these students forward?

Student A

The two diagrams below are equivalent because $\frac{3}{4}$ of each picture is shaded. The second diagram just has more pieces.

Fig. 4.2. Student A's explanation of why $\frac{3}{4}$ is the same amount as $\frac{6}{8}$

Student B

$\frac{6}{8}$, $\frac{9}{12}$, and $\frac{12}{16}$ are all equivalent to $\frac{3}{4}$. I know that all of these fractions are equivalent to $\frac{3}{4}$ because when you divide their numerators and denominators by the same number the answer will become $\frac{3}{4}$. For example, take when 6 is divided by 2 and 8 is divided by 2, the answer is $\frac{3}{4}$. I know to divide the numerator and denominator by the same number because in order for $\frac{3}{4}$ and $\frac{6}{8}$ to be the same, I must be able to divide their numerators and denominators by the same number.

Fig. 4.3. Student B's explanation of why $\frac{3}{4}$ is the same amount as $\frac{6}{8}$

Student C

$\frac{6}{8} = \frac{3}{4}$. Six-eighths is just double three-fourths. They both have the same overall value. If you look at the picture below, you can see the shaded area looks like it is the same in each picture.

Fig. 4.4. Student C's explanation of why $\frac{3}{4}$ is the same amount as $\frac{6}{8}$

Student D

$\frac{6}{8} = \frac{3}{4}$. Six-eighths is the same amount as three-fourths. If you put two of the eighths together, this makes the same as one fourth. Then, since 2 eighths is one fourth, you can make three fourths from the six eighths. See my diagram below.

Fig. 4.5. Student D's explanation of why $\frac{3}{4}$ is the same amount as $\frac{6}{8}$

Which parts of the explanations from students A, B, C, and D did you use to determine whether the thinking was valid? It is useful to consider each explanation in turn.

Student A provides two supporting examples. His unit seems to be the smaller rectangle in the example on the left and the larger rectangle in the example on the right. However, he does not seem to recognize that he must refer to the same unit in situations involving equivalent fractions. In fact, the argument provided by student A seems to invalidate the idea that $\frac{3}{4}$ is equivalent to $\frac{6}{8}$ because the shaded region in the rectangle on the right is much larger than the shaded region in the rectangle on the left. Note that student A is reasoning correctly if he is considering the ratio of the shaded to the unshaded region in each of the two figures, but his reasoning is incorrect in the case of equivalent fractions, which refer to the same whole or unit. (Chapter 7 offers further discussion of the use of ratios and how they differ from fractions.)

Student B provides a lengthy discussion of the procedure that she uses to determine whether two fractions are equivalent. She recognizes that dividing the numerator and the denominator by the same value generates equivalent fractions. However, she does not explain why this procedure works. For example, another student could provide a similar argument and state that adding 2 to the numerator and the denominator generates equivalent fractions. The underlying rationale for this procedure is missing.

Student C makes at least two questionable statements. His first such statement is that $\frac{6}{8}$ is "double" $\frac{3}{4}$. He appears to recognize that the numerator and the denominator are both doubled but seems to be viewing the numerators and the denominators as separate values, rather than recognizing that $\frac{3}{4}$ and $\frac{6}{8}$ represent quantities in and of themselves. In any case, the argument that he should be making is that $\frac{3}{4}$ and $\frac{6}{8}$ are equivalent. We cannot double $\frac{3}{4}$ and arrive at $\frac{6}{8}$ (although many college students do write, incorrectly, that $\frac{3}{4} \times 2 = \frac{6}{8}$). Furthermore, student C states that these amounts "look like" they are the same. However, many amounts may "look" as though they are the same when represented in a diagram (for example, $\frac{99}{100}$ and $\frac{98}{99}$), but this does not mean that they are in fact equivalent.

Student D provides additional information that represents valid support. The reason that $\frac{3}{4}$ and $\frac{6}{8}$ are equivalent has to do with the relationship that exists between fourths and eighths. Because the student knows that she can use 2 eighths to make

1 fourth, she understands that she can use 6 eighths to make the same amount, or value, as 3 fourths.

Helping students recognize how the algorithms for generating equivalent fractions relate to the meaning assigned to physical models is a particular challenge in developing a deep understanding of fraction equivalence. Consider the situation where a student begins with $\frac{2}{3}$ and generates the equivalent fraction $\frac{4}{6}$. If the student has a defined unit and identifies $\frac{2}{3}$ of that unit, then to generate $\frac{4}{6}$, she must partition each of the thirds in half. What happens to the number of pieces that she now has? The number of pieces doubles, since she partitions each third in half, and, at the same time, the number of pieces in the unit also doubles. Similarly, if another student starts with $\frac{2}{3}$ and wants to convert this fraction to the equivalent fraction $\frac{10}{15}$, he needs to partition each third into five smaller pieces, increasing the number of pieces to 5 times as many. This is one valid explanation of why $\frac{2}{3} \times \frac{5}{5}$ equals $\frac{10}{15}$. This is the argument that we typically expect students to use before they have learned how to multiply fractions.

What happens when a student begins with a fraction such as $\frac{8}{10}$? He can simplify this fraction to $\frac{4}{5}$ by combining each pair of tenths to make 1 fifth, thus changing the amount to $\frac{4}{5}$. Note that we use the word "simplify" rather than "reduce"—a term that can be confusing to students, suggesting that the "reduced" fraction is less than its original amount. What happens to the number of partitions that the student has? He decreases the number of partitions by half, or divides the number of partitions by 2.

When introducing students to fraction equivalence, you may have found that sharing situations are natural contexts for helping them develop essential understandings. For example, Empson (1995) describes two typical ways that elementary students determine the fraction of a cake that each child will receive when 20 children share 8 cakes. Some students solved this problem by dividing each cake into 20 equal parts, resulting in each child receiving $\frac{1}{20}$ of each of the 8 cakes, or $\frac{8}{20}$ of a cake. Other students divided each cake into 10 equal parts and gave each child 1 part from each *pair* of cakes, resulting in $\frac{4}{10}$ of a cake for each child. At this point, further discussion occurred in the classroom about which students were correct— the students who said that each child received $\frac{4}{10}$ of a cake, or those who said that each child received $\frac{8}{20}$ of a cake.

At the beginning of their work with equivalent fractions, students typically find situations that involve repeated halving—that is, dividing into halves, fourths, eighths, and so on—more accessible than those that involve dividing into other numbers of equal-sized parts. As students become more familiar with partitioning, they can divide objects into thirds, sixths, and so on.

As noted by Van de Walle (2007), "Many teachers seem to believe that fraction answers are incorrect if not in simplest or lowest terms. This is unfortunate" (p. 312). Requiring that every fraction be written in lowest terms may simplify the grading process, but at the very least, this practice adds an extra, and sometimes extraneous, step to many tasks. In fact, requiring students to write every fraction in lowest terms may cause them to question their understanding of rational numbers or to disregard various equivalent forms of rational numbers. Figure 4.6 shows the response of Lucy, a student in grade 5, to the work of the fictitious student Demetrius, originally shown in figure 2.1. Lucy recognized that $\frac{3}{2}$ is an improper fraction, but she suggested that Demetrius is incorrect because he used an improper fraction. In fact, one-fourth of the students in this class wrote responses similar to Lucy's.

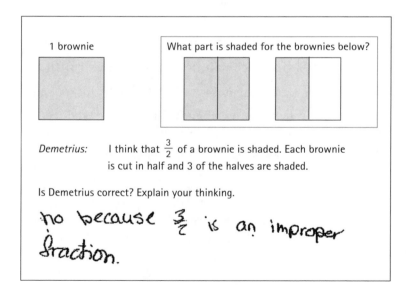

Fig. 4.6. Lucy's response to the work of the fictitious student Demetrius

Figures 4.7 and 4.8 present a pair of tasks that are related to writing numbers that are equivalent to $\frac{1}{2}$. Both tasks were presented to students on the same assessment, about 20 minutes apart. Examine the two tasks while considering the questions in Reflect 4.2.

Reflect 4.2

What aspects of students' understanding of fractions are being assessed in the tasks shown in figures 4.7 and 4.8?

How do you think students in grades 3–5 would respond to these tasks?

Will the responses to the first task differ from the responses from the second task?

If so, how? If not, why?

Write all the fractions or decimals that are equivalent to $\frac{1}{2}$.

Fig. 4.7. A task asking for fractions and decimals equivalent to $\frac{1}{2}$

Ms. Jones asked her students to label the number line below. Write all the fractions that would be correct for where the question mark is on the number line.

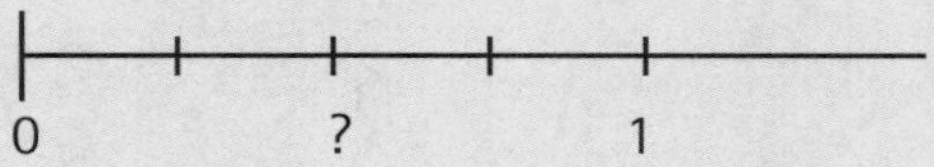

Fig. 4.8. A task asking for labels for a point on the number line

Figures 4.9, 4.10, and 4.11 show responses from three students, Patrick, Cora, and Brandy, respectively, whose work provides some insights into ways that students perceive equivalency. Respond to the questions in Reflect 4.3 as you compare their responses and determine their understandings as well as their misconceptions.

Reflect 4.3

Compare your predictions from Reflect 4.2 with the student responses in figures 4.9, 4.10, and 4.11.

Which students appear to have an understanding of equivalent fractions? What evidence can you cite to support these claims?

What misunderstandings do these students demonstrate related to equivalent fractions?

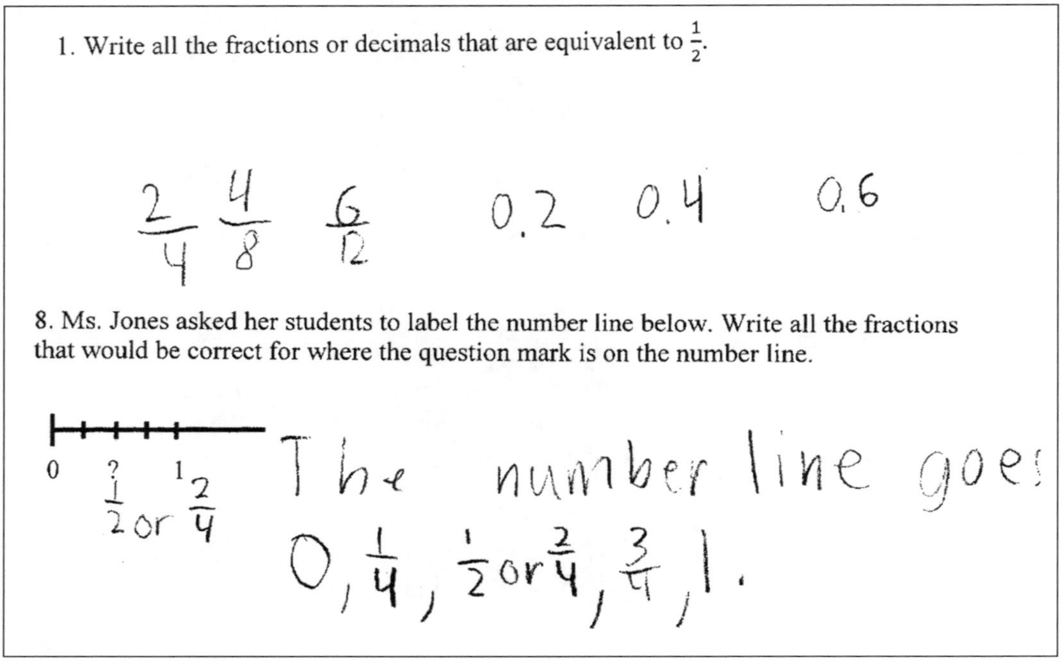

---

1. Write all the fractions or decimals that are equivalent to $\frac{1}{2}$.

$$\frac{2}{4} \quad \frac{4}{8} \quad \frac{6}{12} \qquad 0.2 \quad 0.4 \qquad 0.6$$

8. Ms. Jones asked her students to label the number line below. Write all the fractions that would be correct for where the question mark is on the number line.

The number line goes
$0, \frac{1}{4}, \frac{1}{2}$ or $\frac{2}{4}, \frac{3}{4}, 1$.

Fig. 4.9. Patrick's work on tasks related to fractions and decimals equivalent to $\frac{1}{2}$

1. Write all the fractions or decimals that are equivalent to $\frac{1}{2}$.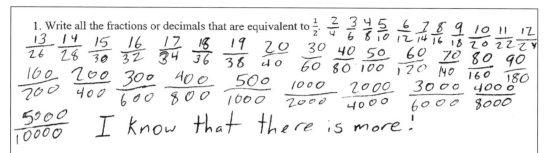

$\frac{2}{4}$ $\frac{3}{6}$ $\frac{4}{8}$ $\frac{5}{10}$ $\frac{6}{12}$ $\frac{7}{14}$ $\frac{8}{16}$ $\frac{9}{18}$ $\frac{10}{20}$ $\frac{11}{22}$ $\frac{12}{24}$

$\frac{13}{26}$ $\frac{14}{28}$ $\frac{15}{30}$ $\frac{16}{32}$ $\frac{17}{34}$ $\frac{18}{36}$ $\frac{19}{38}$ $\frac{20}{40}$ $\frac{30}{60}$ $\frac{40}{80}$ $\frac{50}{100}$ $\frac{60}{120}$ $\frac{70}{140}$ $\frac{80}{160}$ $\frac{90}{180}$

$\frac{100}{200}$ $\frac{200}{400}$ $\frac{300}{600}$ $\frac{400}{800}$ $\frac{500}{1000}$ $\frac{1000}{2000}$ $\frac{2000}{4000}$ $\frac{3000}{6000}$ $\frac{4000}{8000}$

$\frac{5000}{10000}$ I know that there is more!

8. Ms. Jones asked her students to label the number line below. Write all the fractions that would be correct for where the question mark is on the number line.

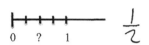 $\frac{1}{2}$

0   ?   1

There can only be 1 fraction on that spot.

Fig. 4.10. Cora's work on tasks related to fractions and decimals equivalent to $\frac{1}{2}$

---

1. Write all the fractions or decimals that are equivalent to $\frac{1}{2}$.

$\frac{2}{4}$ $\frac{3}{6}$ $\frac{4}{8}$ $\frac{5}{10}$ $\frac{6}{12}$ $\frac{7}{14}$ $\frac{8}{16}$ $\frac{9}{18}$ $\frac{10}{20}$ $\frac{11}{22}$ $\frac{12}{24}$ $\frac{13}{26}$ $\frac{14}{28}$ $\frac{15}{30}$ $\frac{16}{32}$ $\frac{17}{34}$ $\frac{18}{36}$ $\frac{19}{38}$

$\frac{20}{40}$ $\frac{21}{42}$ $\frac{22}{44}$ $\frac{23}{46}$ $\frac{24}{48}$ $\frac{25}{50}$ $\frac{30}{60}$ $\frac{40}{80}$ $\frac{50}{100}$ + more

8. Ms. Jones asked her students to label the number line below. Write all the fractions that would be correct for where the question mark is on the number line.

0   ?   1

$\frac{1}{2}$ $\frac{2}{4}$ $\frac{3}{6}$ $\frac{4}{8}$ $\frac{5}{10}$ $\frac{6}{12}$ $\frac{7}{14}$ $\frac{8}{16}$ $\frac{9}{18}$ $\frac{10}{20}$ $\frac{11}{22}$ $\frac{12}{24}$ $\frac{13}{26}$ $\frac{14}{28}$ $\frac{15}{30}$

Fig. 4.11. Brandy's work on tasks related to fractions and decimals equivalent to $\frac{1}{2}$

Clearly, it is not possible to write *all* the fractions and decimals that are equivalent to $\frac{1}{2}$, since the list is infinitely long. Both Brandy and Cora recognized that there are more equivalent fractions than those that they listed. What about Patrick? Can we conclude that he believes that only three fractions are equivalent to $\frac{1}{2}$, or did he just stop listing them? Further, we note that Patrick had some difficulty in representing $\frac{1}{2}$ as a decimal, while Cora and Brandy did not list decimals in their answers. On the task involving the number line, Patrick suggests only two possibilities, whereas Cora insists that only one fraction, $\frac{1}{2}$, is correct. Brandy, by contrast, provided a response that is similar to the one that she gave for the earlier task.

Although we may know some of what Patrick, Cora, and Brandy understand about equivalent fractions, we certainly don't have a complete picture. These tasks may reveal some misunderstandings, but they also generate other questions about what the students do understand. How could we modify these tasks to help us gain more information?

The task shown in figure 4.12 suggests one possibility. Once again, students are asked to respond to the thinking of fictitious students. Their responses highlight a few correct responses as well as a common misconception.

Read the responses of the students to the problem below.

List all fractions that could be located at the question mark below.

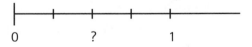

0   ?   1

(a) *Justin:* I think that this spot on the number line is $\frac{1}{2}$.
Is Justin correct? Explain your thinking.

(b) *Elizabeth:* I think that this spot on the number line is $\frac{2}{4}$.
Is Elizabeth correct? Explain your thinking.

(c) *Jane:* I think that this spot on the number line is $\frac{3}{5}$.
Is Jane correct? Explain your thinking.

(d) *Ethan:* I think that this spot on the number line is 0.5.
Is Ethan correct? Explain your thinking.

Fig. 4.12. A task examining fractions equivalent to $\frac{1}{2}$

Comparison of fractions

Barnett-Clarke and colleagues (2010) describe three strategies for comparing fractions, and CCSSM expects students in grades 3 and 4 to understand and use these strategies. It is useful to examine these comparison strategies in greater detail. Each one provides different insights into the relationship between two fractions that are important for students to understand. Students should choose a comparison strategy on the basis of the numbers that they are comparing. Given two fractions, they will often find that one strategy is much more practical than the others to use in making the comparison.

As emphasized earlier, the underlying assumption when comparing two fractions is that the unit that both fractions refer to is the same. For example, when we compare $\frac{3}{4}$ and $\frac{1}{2}$, we can conclude that $\frac{3}{4}$ is greater than $\frac{1}{2}$ as long as the unit—and it could be any unit—is the same. Before exploring the three comparison strategies described by Barnett-Clarke and colleagues, examine the pairs of fractions in Reflect 4.4, and think about the steps that you take to compare the fractions in each pair.

Reflect 4.4

Consider the following fraction pairs. Is one fraction greater than the other? Are the two fractions equivalent? As you consider each example, think about the strategy that you use to compare these fractions.

(a) $\frac{2}{5}$ and $\frac{2}{7}$

(b) $\frac{4}{10}$ and $\frac{3}{8}$

(c) $\frac{6}{11}$ and $\frac{3}{5}$

(d) $\frac{3}{4}$ and $\frac{5}{8}$

(e) $\frac{6}{5}$ and $\frac{8}{7}$

(f) $\frac{8}{17}$ and $\frac{12}{21}$

One of the three strategies for comparing fractions involves obtaining common numerators. Let's take this strategy first because it is useful for comparing the fractions in (a) in our list. Which fraction is greater: $\frac{2}{5}$ or $\frac{2}{7}$? These two fractions have common numerators. If we think about what the denominator means in this situation, we realize that the unit partitioned into sevenths has smaller pieces than that same unit partitioned into fifths. We can reason that $\frac{2}{5}$ is therefore greater than $\frac{2}{7}$. Notice that this reasoning relies on an important idea: the larger the denominator, the smaller the pieces into which the unit is partitioned.

Does our list include other pairs of fractions that we could compare efficiently by using this strategy? Consider the comparison in (c) between $\frac{6}{11}$ and $\frac{3}{5}$. We could easily generate common numerators for this comparison. We could generate the equivalent fraction $\frac{6}{10}$ for $\frac{3}{5}$ and compare $\frac{6}{10}$ and $\frac{6}{11}$. Again, we could reason that $\frac{6}{10}$ is greater than $\frac{6}{11}$ because tenths are larger than elevenths and we have the same number of tenths as we have elevenths. Obtaining a common numerator can be a powerful and efficient strategy for comparing fractions when the fractions have small numerators or one numerator is a multiple of the other.

A second strategy for comparing fractions involves using common denominators. This strategy is often taught in school. The common denominator strategy relies on the fact that if we have parts of the same size, we can simply determine which fraction has more or fewer of these same-sized parts. This strategy is a natural one for use in comparing the fractions in (d): $\frac{3}{4}$ and $\frac{5}{8}$. We can easily generate the equivalent fraction $\frac{6}{8}$ for $\frac{3}{4}$ and compare $\frac{6}{8}$ with $\frac{5}{8}$. However, this strategy can be cumbersome, even for relatively small denominators. Furthermore, this strategy can interfere with the development of a deeper understanding of the meaning of fractions for some students, since they may simply apply the ideas they learned with whole numbers to fractions.

A third strategy for comparing fractions involves using fraction benchmarks. This strategy is one that students often come up with on their own (see Reys, Kim, and Bay [1999]), even when instruction focuses on other strategies. For this strategy, students use a benchmark, such as $\frac{1}{2}$ or 1, to compare fractions. For example, in comparing $\frac{8}{17}$ and $\frac{12}{21}$, the fractions in (f), we might note that $\frac{8}{17}$ is less than $\frac{1}{2}$, since 8.5 is half of 17. By contrast, $\frac{12}{21}$ is greater than $\frac{1}{2}$, since 10.5 is half of 21. Therefore, we could conclude that $\frac{12}{21}$ is greater than $\frac{8}{17}$.

A more complex use of benchmarks involves comparing two fractions that are either both greater than or both less than the benchmark. For example, when comparing $\frac{6}{5}$ and $\frac{8}{7}$, the fractions in (e), we recognize that both fractions are greater than 1, since $\frac{6}{5}$ is $\frac{1}{5}$ more than 1, and $\frac{8}{7}$ is $\frac{1}{7}$ more than 1. Because we know that $\frac{1}{5}$ is greater than $\frac{1}{7}$, we also know that $\frac{6}{5}$ is farther to the right on the number line than $\frac{8}{7}$ and therefore greater than $\frac{8}{7}$. We can apply similar reasoning to fractions that are both less than a benchmark. Consider the pair of fractions in (b), $\frac{4}{10}$ and $\frac{3}{8}$. In this case, both fractions are less than $\frac{1}{2}$. Four-tenths is $\frac{1}{10}$ less than $\frac{1}{2}$, and $\frac{3}{8}$ is $\frac{1}{8}$ less than $\frac{1}{2}$. Because we know that $\frac{1}{10}$ is less than $\frac{1}{8}$, we also know that $\frac{4}{10}$ is closer to $\frac{1}{2}$ on the number line and farther to the right on it than $\frac{3}{8}$. Therefore, $\frac{4}{10}$ is greater than $\frac{3}{8}$.

CCSSM expects students in grade 3 to "compare two fractions with the same numerator or the same denominator by reasoning about their size" (CCSSM 3.NF.3d; NGA Center and CCSSO 2010, p. 24). By the end of grade 4, CCSSM expects students to "compare two fractions with different numerators and different denominators, e.g., by creating common denominators or numerators, or by comparing to a benchmark fraction such as $\frac{1}{2}$" (4NF.2, p. 30).

Note that CCSSM does not describe the use of cross multiplication—a strategy often taught in U.S. elementary classrooms. This strategy can be confusing for students because the underlying reason why it works is often not discussed. Further, this strategy often hinders students from developing a deeper understanding of the meaning of fractions because it does not encourage them to see fractions as numbers but instead focuses their attention on the separate values of the numerators and the denominators. For these reasons, the use of cross multiplication as a strategy is best avoided.

When considering an instructional sequence and selecting instructional tasks, you might find it useful to refer to Bray and Abreu-Sanchez (2010). These classroom teachers describe introducing their third-grade students to comparison of fractions with situations involving unit fractions to help them develop understanding and focus on the relationship between the numerators and the denominators. They began with a pizza context in which one student received $\frac{1}{6}$ of a pizza and another student received $\frac{1}{10}$ of a same-sized pizza. Initially some students thought that $\frac{1}{10}$

was more than $\frac{1}{6}$ because, they reasoned, 10 is more than 6. However, as they began to examine diagrams of the situation, they realized that the pizza divided into tenths had smaller pieces than the pizza divided into sixths. The class continued to examine various situations involving unit fractions, comparing $\frac{1}{4}$ and $\frac{1}{5}$, $\frac{1}{10}$ and $\frac{1}{12}$, and $\frac{1}{20}$ and $\frac{1}{18}$. This work allowed the students to generalize about unit fractions.

Next, Bray and Abreu-Sanchez focused their students' attention on situations involving a common denominator. As could be anticipated, in comparing $\frac{3}{10}$ and $\frac{2}{10}$, some students thought that $\frac{2}{10}$ was greater than $\frac{3}{10}$, reasoning that the smaller the number, the larger the fraction. However, through their work in context, the students soon recognized that $\frac{3}{10}$ was greater than $\frac{2}{10}$.

Note that some textbooks allot a similar amount of instructional time for work with situations involving common denominators as for work with situations involving different denominators (and numerators). The time that students need for making sense of situations with the same denominators is far less than the time that they need for making sense of those with differing denominators.

Bray and Abreu-Sanchez (2010) next focused on assisting students in using fraction benchmarks. They introduced a scenario involving a student who ate $\frac{4}{10}$ of a pizza and another student who ate $\frac{1}{2}$ of a pizza. This situation gave the students an opportunity to recognize that having both a larger numerator and a larger denominator does not always mean having a greater amount. Eventually, Bray and Abreu-Sanchez moved their students to examples that involved fractions on either side of a benchmark fraction (for example, $\frac{3}{7}$ and $\frac{4}{6}$ in relation to $\frac{1}{2}$), and then on to examples of fractions that were both less or both more than a particular benchmark, although, as noted in CCSSM, these strategies should receive closer attention in fourth grade.

Summarizing Pedagogical Content Knowledge to Support Essential Understanding 3a

Teaching the mathematical ideas in this chapter requires specialized knowledge related to the four components presented in the Introduction: learners, curriculum, instructional strategies, and assessment. The four sections that follow summarize some examples of these specialized knowledge bases in relation to Essential Understanding 3a. Although we separate them to highlight their importance, we also recognize that they are connected and support one another.

Knowledge of learners

As discussed earlier, Empson (1995) describes two ways that her students solve sharing situations involving cakes. When students are given the opportunity to make sense of problems, they typically approach problem-solving situations in more than one way. Knowledge of learners involves anticipating possible student responses. However, on some occasions, you are likely to encounter student thinking that is unique or unanticipated. Collecting different student approaches to mathematics problems over time can help you to plan for future instruction.

The examples in this chapter illustrate some erroneous ideas that students often form about fractions. For example, they may think that if someone gives an answer as an improper fraction, it is incorrect (fig. 4.6), or that only one fraction can be on a particular "spot" on the number line (fig. 4.10), or that it is not necessary to refer to the same unit in comparing fractions (fig. 4.2). Knowledge of student misconceptions, in addition to an understanding of accurate student thinking, can help you to anticipate, address, and challenge your students' mathematical thinking.

Knowledge of curriculum

CCSSM encourages the use of the three strategies that we have discussed for comparing fractions. Which of these three strategies do your curricular materials use? Which comparison problems suggest the use of a specific strategy? An analysis of your materials can help you to determine which comparison problems to use and how to sequence them. For example, in the list of paired fractions in Reflect 4.4, problem (a) lends itself to the strategy of comparing numerators, problem (d) invites comparing denominators, and problem (c) is solved efficiently through the use of benchmarks.

As this chapter has described, Bray and Abreu-Sanchez (2010) outline a curricular sequence for comparing fractions, starting with the use of unit fractions. How are unit fractions introduced in your curricular materials? Do your materials have a curricular sequence that makes sense and provides sufficient time for students to develop these understandings? Which models and representations do your materials use to compare fractions?

Knowledge of instructional strategies

Encouraging students to work with physical models can reinforce their understanding of the importance of referring to the same unit when comparing fractions. For example, you might use different-sized candy bars (Snickers Bar, Snickers Miniatures, and Snickers Fun Size, for example) to help students understand the relationship between a fraction such as $\frac{1}{2}$ and the reference unit as they compare $\frac{1}{2}$ of the different bars. As students compare $\frac{1}{2}$ of these different-sized candy bars, they should recognize that they could compare them in different ways (for instance, in terms of length or weight). Students will recognize that $\frac{1}{2}$ of these different bars are not equivalent to one another, since the units differ.

When the attribute used for the unit (for example, length) is not explicit, the language used in curricular tasks and in the classroom can present opportunities as well as challenges—especially for English language learners. "Equivalent" refers to the same value or amount. Thus, assuming that the unit is the same for both fractions, when you talk about equivalent fractions, you should use statements such as, "$\frac{1}{2}$ is equivalent to $\frac{2}{4}$," or "$\frac{1}{2}$ is the same amount as $\frac{2}{4}$," and avoid statements such as, "$\frac{1}{2}$ is the same as $\frac{2}{4}$," or "$\frac{1}{2}$ looks like $\frac{2}{4}$."

Knowledge of assessment

Writing can help students develop mathematical concepts (Shepard 1993), and having your students write about mathematics can provide highly valuable access to their understanding (Silver, Kilpatrick, and Schlesinger 1990; Pugalee 1997). Furthermore, mathematical writing has different genres, including the following described by Marks and Mousley (1990):

- Procedure—writing that tells how something is done

- Description—writing that tells what a particular thing is like

- Report—writing that tells what an entire class of things is like

- Explanation—writing that tells the reason why a judgment has been made

- Exposition—writing that presents arguments about why a thesis has been produced

In examining teaching practice related to equivalent fractions, this chapter presented the work of students A, B, C, and D, who were asked to write why $\frac{3}{4}$ is the same amount as $\frac{6}{8}$. This assessment task differs from a task such as, "Show that $\frac{3}{4}$ is the same amount as $\frac{6}{8}$." It is important to help students develop writing competencies that are analogous to those that mathematicians use. Therefore, assessment tasks should require students not only to describe how they solved mathematics problems, but also to provide mathematical arguments, justify their thinking, and generalize beyond a small number of cases. Useful assessment questions include the following:

- Why does this procedure work?

- Will this strategy always work? and

- How do you know?

These questions will not only help your students develop skill with different genres of mathematics writing, but also provide opportunities for you to assess their development of competency in using these genres.

Conclusion

As this chapter has demonstrated, helping your students develop essential understandings related to equivalent fractions and the comparison of fractions is challenging. It requires focusing on meaning and underlying ideas, careful attention to language, purposeful selection of tasks, analysis of student thinking, and thoughtful questions. Carefully selecting tasks and posing effective questions will help students develop these essential understandings and lay the foundation for an essential understanding of decimal fractions, as we discuss in the next chapter.

into

practice

Chapter 5
Decimal Fractions

Essential Understanding 3c
A rational number can be expressed as a decimal.

Mathematics curricula in the United States typically introduce fractions before decimals, or decimal fractions. It is important that students connect the meaning that they have learned for fractions with the meaning that they begin to understand for decimals. *Developing Essential Understanding of Rational Numbers for Teaching Mathematics in Grades 3–5* (Barnett-Clarke et al. 2010) presents the idea of expressing a rational number as a decimal fraction as Essential Understanding 3c. Helping students develop an understanding of this idea is the focus of this chapter. Providing opportunities for students to represent particular fractions as decimals and particular decimals as fractions can assist them in developing these connections.

Working toward Essential Understanding 3c

The Common Core State Standards for Mathematics (CCSSM; National Governors Association and Council of Chief State School Officers [NGA Center and CCSSO] 2010) call for attention to decimals in grade 4. CCSSM expects students at this level to "understand decimal notation for fractions, and compare decimal fractions" (4.NF, p. 31), recommending that fourth graders use decimal notation to represent fractions with a denominator of 10 or 100 (see fig. 5.1).

Common Core State Standards for Mathematics, Grade 4

Understand decimal notation for fractions, and compare decimal fractions.

5. Express a fraction with denominator 10 as an equivalent fraction with denominator 100, and use this technique to add two fractions with respective denominators 10 and 100.

   For example, express $\frac{3}{10}$ as $\frac{30}{100}$, and add $\frac{3}{10} + \frac{4}{100} = \frac{34}{100}$.

6. Use decimal notation for fractions with denominators 10 or 100.

   For example, rewrite 0.62 as $\frac{62}{100}$; describe a length as 0.62 meters; locate 0.62 on a number line diagram.

7. Compare two decimals to hundredths by reasoning about their size. Recognize that comparisons are valid only when the two decimals refer to the same whole. Record the results of comparisons with the symbols >, =, or <, and justify the conclusions, e.g., by using a visual model.

Fig. 5.1. Decimal notation, CCSSM 4.NF.5–7 (NGA Center and CCSSO 2010, p. 31)

In grades 4 and 5, students can investigate fractions such as halves, fourths, and fifths, which can easily be represented as decimals. Providing opportunities for students to share decimals and relate them to fractions with which they are familiar can help them develop the understanding called for in CCSSM. Students may encounter confusion as they confront the fact that rational numbers can be represented not only as various fractions (for example, $\frac{1}{2}$, $\frac{3}{6}$, $\frac{5}{10}$) that do not look the same but are equivalent, but also as various decimals (0.5, 0.50, 0.500) that likewise look different but are equivalent.

The task in figure 5.2 could be used to help students connect the meaning of fractions and decimals with a physical representation of the amount. This task requires students to shade 0.6 of the 10-by-10 grid and to determine other fractions that could be used to represent the shaded amount.

A caution that we stressed for working with students on fractions is also relevant for working with them on decimal fractions—the unit must be clearly established. One of the misconceptions that can develop when students are introduced to decimal notation involves the failure to recognize the suffix "-ths" at the end of "tenths" and "hundredths." It is important that students recognize $\frac{1}{10}$ and 0.1 and see both as representing one tenth of a unit, or whole.

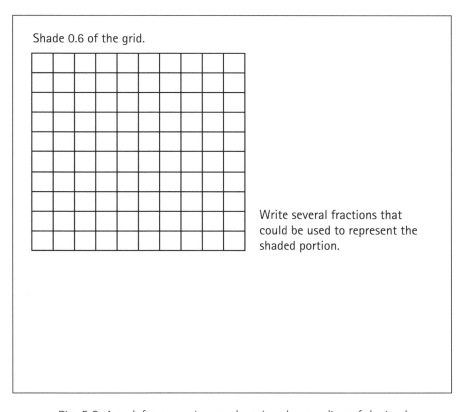

Shade 0.6 of the grid.

Write several fractions that could be used to represent the shaded portion.

Fig. 5.2. A task for assessing students' understanding of decimals

Just as in representing fractions, students should be asked to represent decimals by using length, area, and discrete models. A misconception that often surfaced in our work with students in fourth and fifth grade was the idea that only fractions, and not decimals, can be placed on the number line. Figure 5.3 shows the work of a fourth-grade student with this misunderstanding.

Periodically varying the unit when representing decimals and fractions can reveal students' misconceptions about the unit. For example, the task in figure 5.4 offers a follow-up to the task in figure 5.2. Again, students are asked to shade six-tenths of a given grid. However, as a result of the change in unit, students must think carefully about what $\frac{1}{10}$ of the unit is. Students who are confused and view 0.6 as "six tens" may shade the entire whole. Such activities can serve as entry points for discussions that can clarify students' misconceptions about the meaning of decimal notation.

Two students are trying to determine what number goes where the question mark is between 0 and 1 on the number line.

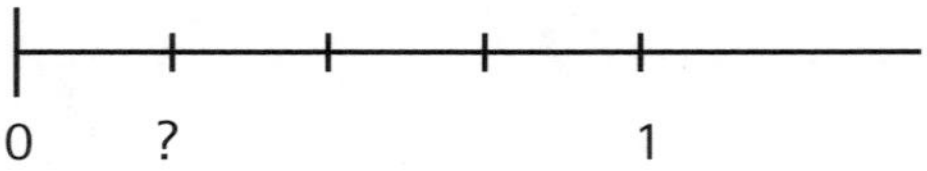

| | | | | | | |
| | | | | | | |

0   ?   1

Parker: This spot on the number line is where 0.25 would go.

Is Parker correct? Explain your thinking.

NO, because he is saying it in desmole form.

Brandon: This spot on the number line is where $\frac{1}{4}$ would go.

Is Brandon correct? Explain your thinking.

Yes, because he is saying it in frauton form.

Fig. 5.3. Work by a fourth-grade student, indicating the misconception that decimals may not be placed on a number line

As students continue to develop an understanding of the meaning of decimal fractions, they can deepen their understanding by comparing decimals. CCSSM expects students in fourth grade to "compare two decimals to hundredths by reasoning about their size" (4.NF.7; NGA Center and CCSSO 2010, p. 31). Research reveals that students and adults alike hold various misconceptions about decimals that surface when they are asked to compare decimals (Widjaja, Stacey, and Steinle 2011). The following is a short list of misconceptions that students and adults often demonstrate when comparing decimals:

- Longer is larger. The longer the decimal, the larger the number. Students with this misconception may state, for example, that 0.38 is larger than 0.6.

- Zeros can be ignored. "Zero" has no value and does not affect the size of the number. Students with this misconception may state, for example, that 4.03 and 4.3 are equal.

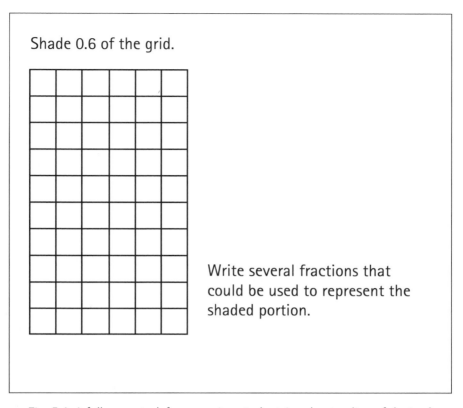

Shade 0.6 of the grid.

Write several fractions that could be used to represent the shaded portion.

Fig. 5.4. A follow-up task for assessing students' understanding of decimals

- The decimal equals a "reciprocal." Students with this misconception are confused about the meaning of the numerator and denominator for decimal fractions, and they may, for example, interpret 0.3 as $\frac{1}{3}$ and 0.4 as $\frac{1}{4}$ and state that 0.3 is greater than 0.4.

Because these misconceptions emerge in students' work with decimals, it is important that tasks be carefully designed to reveal them. Further, it is important that students state how they determined that one decimal is greater or less than another decimal. Tasks in many textbooks that we examined did not meet either of these recommendations and would provide teachers with little insight to address student misconceptions.

For their misconceptions to become evident, students should be asked to compare decimals in tasks such as that in figure 5.5, providing explanations to support their reasoning. Students should support their responses with diagrams or write arguments, and they should clearly define the unit. Having other students examine valid and invalid arguments can also assist students in identifying and overcoming common misconceptions.

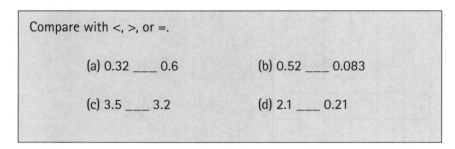

Fig. 5.5. A comparison task with the potential to reveal students' misconceptions

It may be tempting to show students a simple rule, such as, "Add zeros to the end of the decimals to have the same number of digits to the right of the decimal points." However, such instructional strategies often limit students' understanding, since they tend to shift the focus away from the meaning of decimals and toward rules that produce correct answers without building understanding. The misconceptions in our short list were found in adults, demonstrating how persistent student misconceptions can be if they are not identified and confronted.

Summarizing Pedagogical Content Knowledge to Support Essential Understanding 3c

Teaching the mathematical ideas in this chapter requires specialized knowledge related to the four components presented in the Introduction: learners, curriculum, instructional strategies, and assessment. The four sections that follow summarize some examples of these specialized knowledge bases in relation to Essential Understanding 3c. Although we separate them to highlight their importance, we also recognize that they are connected and support one another.

Knowledge of learners

How do your students interpret the length of different decimals? Do they believe that "longer" decimals are always larger? How would your students answer the question, "When are 'longer' decimals smaller than 'shorter' decimals?" (Of course, you would need to ensure that your students understand what you mean by "longer" and "shorter" in this case.) How do your students interpret the value of zeros in different places? When they encounter answers such as $0.50, do they insist that those zeros are not necessary or should be eliminated? How do they interpret a comparison between decimals such as 0.05 and 0.50? Knowledge of student misconceptions, in addition to accurate student thinking, can help you anticipate, address, and challenge students mathematically.

Knowledge of curriculum

CCSSM (2010) stresses that students should reason about the size of decimals. This emphasis is supported by research that highlights common misconceptions that surface when students compare decimals. Therefore, curricular materials should include mathematical tasks that will reveal students' misconceptions as well as require students to reason about decimals. Analyze your curricular materials to determine which decimal comparison problems to use and how to sequence them. How do your materials introduce decimals? Do your materials have a curricular sequence that makes sense and provides sufficient time for students to develop these understandings? Which models and representations do your materials use to introduce decimals and enable students to compare them?

Knowledge of instructional strategies

Capitalizing on students' prior experiences and knowledge is an important instructional strategy. What student experiences can you draw on in teaching about decimals? When have your students encountered decimals? Consider, for example, measurement or money contexts that you might use.

It can also be helpful to connect students' prior experiences with fractions with their new work with decimals. Talking about different notations for the same value—for instance, $\frac{1}{5}$ and 0.20—and discussing when to use different notations may help students continue to build their understanding of equivalency.

Knowledge of assessment

The fourth-grade student whose work is shown in figure 5.3 indicated that decimals could not be represented on a number line. Just as in work with fractions, students should be asked to represent decimals by using length, area, and discrete models.

Moreover, as a teacher, you should carefully consider the unit that you use in assessment tasks. For example, figures 5.2 and 5.4 show tasks that ask students to shade 0.6 of given grids and write several fractions that represent the shaded portions. Including tasks that use different units will not only provide opportunities for you to assess your students' understanding of the essential idea of unit, but also provide your students with opportunities to analyze and compare similar tasks.

Conclusion

Helping your students develop a deep understanding of decimal fractions requires giving them opportunities to consider and work with different models (area, linear, and discrete) as well as both pictorial and symbolic representations. Although these models and representations can support appropriate understandings, they can also lead to misunderstandings—for example, the idea that decimals are not used to label number lines (see fig 5.3). By carefully selecting tasks and posing effective questions, you can help your students develop understandings related to the meaning and comparison of decimal fractions and make connections with other ideas about fractions discussed earlier, such as equivalence, and with ideas involved in computing with fractions, the topic of the next chapter.

Chapter 6
Addition and Subtraction
with Fractions

Big Idea 4

Computation with rational numbers is an extension of computation with whole numbers but introduces some new ideas and processes.

In *Developing Essential Understanding of Rational Numbers for Teaching Mathematics in Grades 3–5* (Barnett-Clarke et al. 2010), addition and subtraction with fractions are the focus of Big Idea 4. This big idea focuses on adding and subtracting fractions as extensions of adding and subtracting with whole numbers, although computations with fractions incorporate new ideas and processes. Thus, the essential understandings that Barnett-Clarke and colleagues identify in relation to adding and subtracting fractions are rooted in the essential understandings for adding and subtracting whole numbers identified in *Developing Essential Understanding of Addition and Subtraction for Teaching Mathematics in Prekindergarten–Grade 2* (Caldwell, Karp, and Bay-Williams 2011). The development of students' understanding of these interrelated ideas and processes is the focus of this chapter.

Working toward Big Idea 4

Children in the early grades need to understand that "many different problem situations can be represented by ... addition and subtraction" and "the context of a problem situation and its interpretation can lead to different representations" (Caldwell, Karp, and Bay-Williams 2010, p. 10, Essential Understandings 1c and 1e). The inclusion of rational numbers gives these operations a new layer of complexity. In particular, the unit serves a critical role in work with fractions and mixed numbers. This chapter classifies problem situations involving addition and subtraction and discusses models that can be used to represent these situations. It begins, however, with a discussion of the sequencing of addition and subtraction problems involving fractions.

Sequencing addition and subtraction problems

Mathematics textbooks for students in grades 3–5 typically devote a single lesson to adding and subtracting fractions with like denominators, followed by a single lesson on adding and subtracting fractions with unlike denominators. In subsequent lessons, problems with like denominators are mixed among those with unlike denominators, and the numbers gradually increase in size from those that are less than 1 to mixed numbers or improper fractions.

It is easy to see how students moving at a pace of one lesson a day become confused about the process of adding and subtracting fractions. If students have a deep understanding of the meaning of fractions, adding fractions with like denominators will not require very much time (probably less than one class period), but developing a deep understanding of adding and subtracting with unlike denominators is likely to require more than a week.

Mack (2004) suggests a variation on the traditional sequence of fraction problems, beginning with addition and subtraction of fractions with like denominators, and gradually incorporating whole numbers or mixed numbers with like denominators. She recommends posing problems in a context and suggests including fractions that are less than 1 along with whole numbers and mixed numbers. Students might be offered problems such as the following:

- Gerald ate $\frac{3}{12}$ of a cherry pie, and Donna ate $\frac{1}{12}$ of the same cherry pie. How much of the cherry pie did Gerald and Donna eat?

- Marty had 4 doughnuts. He ate $\frac{3}{4}$ of one doughnut. How many doughnuts does Marty have now?

- Thomas had $2\frac{1}{4}$ pounds of meat. He used $\frac{3}{4}$ of a pound to make a meatloaf. How much meat does Thomas have left?

Students can use concrete and pictorial models to represent these situations and share their solution strategies with the class. Once they are comfortable with working with given numbers of like-sized pieces, Mack suggests offering them problems involving fractions or mixed numbers with different denominators, beginning with one denominator that is a multiple of the other. Students may realize that they can rename one of the given fractions as an equivalent fraction, again obtaining particular numbers of like-sized pieces to add or subtract. Finally, Mack suggests using fractions with different denominators that do not share a common factor, such as 9 and 4. This ensures that the least common multiple of the denominators is the product of the two denominators, as in $\frac{8}{9} + \frac{3}{4} = \frac{32}{36} + \frac{27}{36} = \frac{59}{36}$, or $1\frac{23}{36}$.

Following this sequence of problems, Mack suggests having students focus on adding or subtracting given numbers of like-sized pieces and using equivalent fractions when necessary. This helps avoid the common error of adding both the numerators and the denominators. For example, one student, Greg, gave the solution shown in figure 6.1 when Mack posed the following problem: "You have three-eighths of a medium pepperoni pizza. I give you two-eighths more of a medium pepperoni pizza. How much of a medium pepperoni pizza do you have now?" (p. 229).

Figure 2

Greg's solution to problem corresponding to "3/8 + 2/8 = ?"

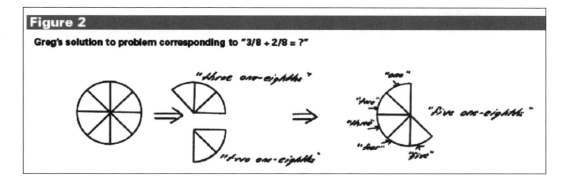

Fig. 6.1. Greg's representation of $\frac{3}{8} + \frac{2}{8}$ (Mack 2004, p. 229)

When Mack asked what students thought about the possibility of $\frac{5}{16}$ as a solution, they were confused about why the denominators would be added together. Another student, Brian, commented on why this answer was not correct:

> Because these are eighths [*holds up one-eighth of a fraction circle*]. If you put them together you still have eighths [*shows this in a manner similar to Greg's*]. See, you didn't make them into sixteenths when you put them together. They're still eighths. (p. 229)

Importance of the unit in addition and subtraction

Addition with fractions, like addition with whole numbers, involves combining disjoint sets. In adding two fractions, it is critical to recognize that both fraction addends need to have the same unit and that the resulting sum will also have that unit. It does not make sense, for example, to use $\frac{1}{2} + \frac{1}{3}$ to find the total length of two ribbons with individual lengths of $\frac{1}{2}$ of a foot and $\frac{1}{3}$ of a yard. To find the total length, we need to use the same unit for both ribbons (perhaps feet, inches, or yards).

The context of a problem affects the way in which someone solves it or interprets an answer. Reflect 6.1 poses questions about the impact of the context of the problems shown in figure 6.2, all of which involve $\frac{1}{2}$ and $\frac{1}{3}$, to help you consider your thinking about a solution process and the answer.

Reflect 6.1

Which problem or problems in figure 6.2 would be solved by adding $\frac{1}{2} + \frac{1}{3}$?

What are the important characteristics of problems that can be solved by adding the fractions mentioned in the problem?

Even when two fractions refer to the same unit, adding or subtracting them may or may not make sense. Of the four problems in figure 6.2, only the first would be solved by adding $\frac{1}{2} + \frac{1}{3}$. In problem 2, the unit for $\frac{1}{2}$ is a cup, but the unit for $\frac{1}{3}$ is "the water in the glass"—after $\frac{1}{2}$ cup has been removed. The two fractions in problem 3 also have different units. In problem 4, the two sets of children may not be disjoint; there may be students who would like to visit both the zoo and the science museum, and adding $\frac{1}{2} + \frac{1}{3}$ would count those students twice.

Let's examine problem 3 in more detail. Suppose that we had more information—perhaps that the number of boys in the class was equal to the number of girls in the class. Could we then solve the problem with $\frac{1}{2} + \frac{1}{3}$? Take, for example, a class of 12 boys and 12 girls. In this case, 6 boys and 4 girls (or 10 students) would be wearing tennis shoes. That would be $\frac{10}{24}$, or $\frac{5}{12}$, of the students in the class,

1. Ari pours $\frac{1}{2}$ of a cup of sand into an empty box. Then Ari pours $\frac{1}{3}$ of a cup of sand into the box. How many cups of sand are in the box now?

2. Tom has a full glass of water. Tom pours $\frac{1}{2}$ cup of water from the glass into an empty bowl. Then Tom pours in another $\frac{1}{3}$ of the water from the glass into the bowl. How many cups of water are in the bowl now?

3. $\frac{1}{2}$ of the boys in the class are wearing tennis shoes. $\frac{1}{3}$ of the girls in the class are wearing tennis shoes. What fraction of the class is wearing tennis shoes?

4. $\frac{1}{2}$ of the children at Russell Elementary School say they would like to visit the zoo. $\frac{1}{3}$ of the children at Russell Elementary School say they would like to visit the science museum. What fraction of the children at Russell Elementary School would like to visit the zoo or the science museum?

Fig. 6.2. Problems involving $\frac{1}{2}$ and $\frac{1}{3}$

whereas $\frac{1}{2} + \frac{1}{3} = \frac{5}{6}$. It is true that 10 students is $\frac{5}{6}$ of 12 students (the unit for both $\frac{1}{2}$ and $\frac{1}{3}$), but we were looking for the fraction of the *class*.

These sample problems point to the importance of referring to the same unit in each of the addends and in the sum when adding two fractions. However, such clear reference to the unit is often absent from discussions of fraction addition, leading students to develop misconceptions about when to use the procedures for adding fractions that they learn in school. As shown in figure 6.3, the Common Core State Standards for Mathematics (CCSSM; National Governors Association Center for Best Practices and Council of Chief State School Officers [NGA Center and CCSSO] 2010) expect students in grade 4 to develop an understanding of the idea that in the addition of fractions, the addends must refer to the same unit, or whole.

As Barnett-Clarke and colleagues (2010) note, students may apply their knowledge of operations with whole numbers inappropriately when working with fractions. The most common mistake that students make when adding fractions is to add the numerators and then the denominators. Helping students develop an understanding of the meaning of the numerators and denominators will help them understand that this approach does not work, since the denominator indicates the number of parts that make a unit.

Common Core State Standards for Mathematics, Grade 4

Build fractions from unit fractions by applying and extending previous understandings of operations on whole numbers.

3. Understand a fraction $\frac{a}{b}$ with $a > 1$ as a sum of fractions $\frac{1}{b}$.

    a. Understand addition and subtraction of fractions as joining and separating parts referring to the same whole.

Fig. 6.3. Addition and subtraction of fractions, CCSSM 4.NF.3a
(NGA Center and CCSSO 2010, p. 30)

Are there situations in which it is appropriate to add numerators and denominators? Respond to the questions in Reflect 6.2 to examine Kevin's addition of fractions in the context of free throw shots, as shown in figure 6.4.

Reflect 6.2

In figure 6.4, Kevin uses fractions to analyze his success in making free throw shots in two basketball games. Is Kevin correct in his thinking?

If so, why does Kevin's method work? If not, what does Kevin not understand?

Kevin plays on a basketball team. In class one day, he says, "I made 3 of 5 free throws in one game and 2 of 6 free throws in another game. Therefore, I made $\frac{3}{5} + \frac{2}{6} = \frac{5}{11}$, or $\frac{5}{11}$, of my free throws. When you add fractions, you should add the numerators and denominators."

Fig. 6.4. Kevin's addition of numerators and denominators to report his free-throw success in two basketball games

Kevin is correct that he made $\frac{3}{5}$ of his free throws in the first game, $\frac{2}{6}$ in the second game, and $\frac{5}{11}$ in the two games. The units for the two fractions are actually different from each other, and different from the unit of the sum. Yet, adding numerators and denominators "works," for a subtle reason. In fact, the $\frac{3}{5}$, $\frac{2}{6}$, and $\frac{5}{11}$ are all *ratios* that refer to different units, and are dealt with in a slightly different manner from fractions. (Chapter 7 provides more information about ratios.)

The problems in figure 6.5 resemble those in figure 6.2, with the new set involving the fractions $\frac{2}{3}$ and $\frac{1}{2}$. Read the problems while thinking about the questions in Reflect 6.3.

Reflect 6.3

Which problems in figure 6.5 can be solved by subtracting $\frac{2}{3} - \frac{1}{2}$?

What are the important characteristics of problems that can be solved by subtracting one fraction from another?

---

1. $\frac{2}{3}$ of the kingdom of Reyna is forestland. $\frac{1}{2}$ of the neighboring kingdom of Lexia is forestland. How much more forestland is there in Reyna than in Lexia?

2. $\frac{2}{3}$ of the children at Lincoln Elementary rode the Ferris wheel at the county fair. $\frac{1}{2}$ of the children at Lincoln Elementary rode the carousel. What fraction of children at Lincoln Elementary rode the Ferris wheel, but not the carousel?

3. Starting at her house, Katie bikes $\frac{2}{3}$ of a mile down the street. Then Katie turns around and bikes $\frac{1}{2}$ mile back toward her house. How far down the street is Katie from her house?

4. Brett pours $\frac{2}{3}$ cup of water into his mug. Then Brett pours out $\frac{1}{2}$ of the water that is in his mug into the sink. How many cups of water are in Brett's mug now?

Fig. 6.5. Problems involving $\frac{2}{3}$ and $\frac{1}{2}$

Like fractions that are added together, fractions that are subtracted from one another must refer to the same unit. Even so, this does not guarantee that subtracting one fraction from another in a problem will produce the desired result. Of the four problems in figure 6.5, only problem 3 is solved by computing $\frac{2}{3} - \frac{1}{2}$.

It is not appropriate to compute $\frac{2}{3} - \frac{1}{2}$ to solve the other three problems. In problem 1, the units for the two fractions may be different, since we do not know the sizes of the two kingdoms. Further, the problem is asking for an actual amount of forest-land, not a fraction. The two fractions in problem 2 have the same unit, but there may be students who rode both rides, or only the carousel, or only the Ferris wheel. That is, the children riding the carousel may not form a subset of the children who rode the Ferris wheel. In problem 4, the units are different; $\frac{2}{3}$ refers to a cup of water, while $\frac{1}{2}$ refers to the water in the mug. Therefore, the unit for $\frac{1}{2}$ is in fact $\frac{2}{3}$ of a cup. Problem 4 can actually be solved by multiplication, using the fraction $\frac{1}{2}$ as an operator. (See Barnett-Clarke and colleagues [2010] for a discussion of fraction as operator, and *Putting Essential Understanding of Multiplication and Division into Practice in Grades 3–5* [forthcoming] for a discussion of multiplication of fractions.) Like the addends and sum in the addition of fractions, the minuend, subtrahend, and difference must all refer to the same unit in the subtraction of fractions. This is an idea that CCSSM expects students in grade 4 to understand (see fig. 6.6).

Common Core State Standards for Mathematics, Grade 4

Build fractions from unit fractions by applying and extending previous understandings of operations on whole numbers.

3. Understand a fraction $\frac{a}{b}$ with $a > 1$ as a sum of fractions $\frac{1}{b}$.

    c. Add and subtract mixed numbers with like denominators, e.g., by replacing each mixed number with an equivalent fraction, and/or by using properties of operations and the relationship between addition and subtraction.

    d. Solve word problems involving addition and subtraction of fractions referring to the same whole and having like denominators, e.g., by using visual fraction models and equations to represent the problem.

Fig. 6.6. Addition and subtraction of fractions, CCSSM 4.NF.3c and 3d
(NGA Center and CCSSO 2010, p. 30)

Working with addition and subtraction situations

Addition and subtraction situations can be divided into four basic classes, described by Carpenter, Fennema, and Franke (1996) as problems involving: "(a) joining action, (b) separating action, (c) part-part-whole relations, and (d) comparison situations" (p. 6). Students may solve a so-called subtraction problem by using addition. Take, for example, the problem shown in figure 6.7, which Reflect 6.4 invites you to view in different ways.

Reflect 6.4

How can addition be used to solve the problem in figure 6.7?

How can subtraction be used to solve the problem?

Johanna needs $3\frac{1}{2}$ cups of flour for a cake. She only has $2\frac{1}{4}$ cups of flour. How much more flour does Johanna need to make the cake?

Fig. 6.7. A "subtraction problem" that is easily solved by addition

Of the four classes of addition and subtraction problems identified by Carpenter, Fennema, and Franke (1996), the problem in figure 6.7 would best be described as a part-part-whole problem. One part is $2\frac{1}{4}$ cups, the other part is unknown, and the whole is $3\frac{1}{2}$ cups.

Children use different strategies when adding whole numbers (Carpenter, Fennema, and Franke 1996; Fuson 1992; National Research Council 2001). These strategies represent increasing levels of sophistication, as follows:

- Direct modeling. Children use physical objects or fingers to represent both quantities and physically push them together.

- Counting strategies. Children do not represent both of the quantities, but may count on from the first number or count on from the larger number (mentally or physically) to find the sum.

- Derived facts. Children use number facts that they already know, such as doubles, to solve other problems.

In the same way, when students begin to add or subtract fractions, they engage in direct modeling by using physical models or pictorial diagrams. The context and any associated actions may motivate them to make different diagrams, even when the fractions used in the problems are the same. Respond to the questions in Reflect 6.5 as you examine the problems involving $2\frac{1}{2}$ and $\frac{3}{4}$ shown in figure 6.8.

Reflect 6.5

Make a diagram to represent each problem in figure 6.8. How are your diagrams similar? How are they different?

How do the diagrams help reveal the class of addition and subtraction problem (join, separate, part–part–whole, or comparison) to which each problem belongs?

1. Jerry has $2\frac{1}{2}$ pages full of stickers. He gives $\frac{3}{4}$ of a page to his sister Rosalie. How many pages of stickers does Jerry have now?

2. Jerry has $2\frac{1}{2}$ pages full of stickers. His sister Rosalie has $\frac{3}{4}$ of a page of stickers. How many more pages of stickers does Jerry have than Rosalie?

3. Jerry wants to fill $2\frac{1}{2}$ pages in his book with stickers. Right now he has enough stickers to fill $\frac{3}{4}$ of a page. How many more pages full of stickers does Jerry need?

4. Jerry has $2\frac{1}{2}$ pages full of superhero stickers. He has $\frac{3}{4}$ of a page of sports stickers. How many pages of stickers does Jerry have altogether?

Fig. 6.8. Problems involving $2\frac{1}{2}$ and $\frac{3}{4}$

Although each of the problems in figure 6.8 uses the same numbers and similar contexts, they fall into different classifications. Problem 1 involves a separating action—removing $\frac{3}{4}$ of a page of stickers from a larger group of $2\frac{1}{2}$ pages. In problem 2, two different groups are being compared. Problem 3 involves part-part-whole relationships, in which one part is $\frac{3}{4}$ of a page, another part is unknown, and the whole is $2\frac{1}{2}$ pages. Finally, problem 4 involves a joining action, combining two separate groups.

Depending on the context of the situation, different models may be more appropriate, or may make more sense to students, than other models. For example, consider problem 1 in figure 6.2. Ari combines $\frac{1}{2}$ of a cup of sand and $\frac{1}{3}$ of a cup of sand in a box. Figure 6.9 represents this scenario with a rectangular area model. The large rectangle represents the unit, one cup of sand, while the two addends are shaded in different colors. Forming equivalent fractions gives us two fractions with the same denominator. Sixths provide a convenient denominator, allowing us to combine $\frac{3}{6}$ of a cup of sand and $\frac{2}{6}$ of a cup of sand.

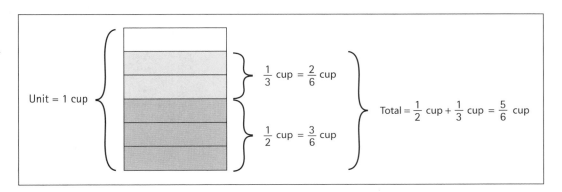

Fig. 6.9. Representing $\frac{1}{2} + \frac{1}{3}$ with a rectangular area model

In problem 3 in figure 6.5, Katie is riding her bike. She bikes $\frac{2}{3}$ of a mile from her house and then turns around and bikes $\frac{1}{2}$ of a mile back. A number line model can represent the action, as shown in figure 6.10. Using equivalent fractions is helpful in interpreting the problem itself and the final answer—that when Katie stops, she is $\frac{1}{6}$ of a mile away from her house.

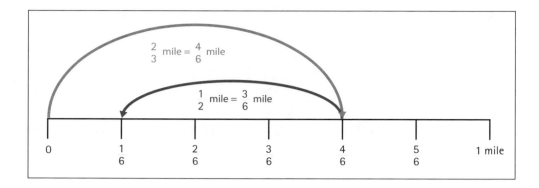

Fig. 6.10. A number line model for $\frac{2}{3} - \frac{1}{2}$

Using direct modeling may help students to represent a problem but may not prevent them from making errors in solving it. The questions in Reflect 6.6 ask you to examine the solutions to a fraction problem offered by five students, as shown in figure 6.11. Consider the representations and reasoning that the students provide. (As we noted in Chapter 3, pattern blocks are usually manufactured in particular standard colors; student B refers to the red trapezoid [with area equal to $\frac{1}{2}$ that of the hexagon] and the blue rhombus [with area equal to $\frac{1}{3}$ that of the hexagon]. Unfortunately, the figure here cannot reproduce the colors that the student names.)

Reflect 6.6

Figure 6.11 shows the reasoning of students A–E as they model a problem involving the addition of $\frac{3}{4}$ and $\frac{2}{3}$.

Which students display errors in their reasoning?

What questions would you ask these students to help them understand their errors?

How could you use the students' diagrams to help them find the sum?

Erin has $\frac{3}{4}$ of a pound of chocolate candies and $\frac{2}{3}$ of a pound of mint candies. How many pounds of candy does she have altogether?

Student A:	I used fraction circles for the $\frac{3}{4}$ and $\frac{2}{3}$. Because the pieces are different sizes, they can't be added together. Also, if you could put them together, the result would be more than one circle.

Student B:	I used pattern blocks with 2 hexagons as the whole, so this

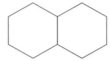

represents 1. That means $\frac{3}{4}$ is 3 red trapezoids and $\frac{2}{3}$ is 4 blue rhombuses. So $\frac{3}{4} + \frac{2}{3} = \frac{7}{10}$.

 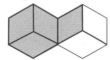

Erin has $\frac{7}{10}$ of a pound of candy.

Student C:	I used rectangles as a whole, and I divided each rectangle into 12 equal pieces.

 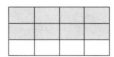

So $\frac{3}{4} + \frac{2}{3} = \frac{17}{24}$. Erin has $\frac{17}{24}$ of a pound of candy.

Student D:	I know that 3 and 4 both multiply to 12, so I replaced $\frac{3}{4}$ with $\frac{9}{12}$ and $\frac{2}{3}$ with $\frac{8}{12}$. Then $\frac{3}{4} + \frac{2}{3}$ becomes $\frac{9}{12} + \frac{8}{12}$. Take $\frac{3}{12}$ from the $\frac{8}{12}$ and put it with the $\frac{9}{12}$ to make $\frac{12}{12}$. You still have $\frac{5}{12}$ left from the $\frac{8}{12}$. So $\frac{3}{4} + \frac{2}{3} = \frac{12}{12} + \frac{5}{12} = 1\frac{5}{12}$. Erin has $1\frac{5}{12}$ pounds of candy.

Student E:	Both of these fractions are a little less than 1, so the sum will be less than 2. Take 2 minus $\frac{1}{4}$, which leaves $1\frac{3}{4}$. Now take away $\frac{1}{3}$. $1\frac{3}{4} - \frac{1}{3}$ is the same as 1.75 − .333... = 1.4666.... So Erin has 1.4666... pounds of candy.

Fig. 6.11. Five students' answers to $\frac{3}{4} + \frac{2}{3}$

Student A models the two amounts by using circles but is hindered by the fact that the pieces appear to be different sizes, as well as by the fact that the result would be greater than 1. Student B also uses an area model, but views the pieces discretely. Furthermore, this student abandons the idea of two hexagons as a unit and bases his answer on four hexagons as a unit. Student C makes a similar error, switching the unit during the process from one rectangle to two rectangles. This error is a common one for students to make when the fractions sum to a number larger than 1. Students D and E both give correct answers, and both use derived facts to determine the sum. As shown in figure 6.12, CCSSM expects students in grade 4 to be able to decompose a fraction into a sum of fractions with the same denominator, a skill that supports the reasoning of students D and E.

Common Core State Standards for Mathematics, Grade 4

Build fractions from unit fractions by applying and extending previous understandings of operations on whole numbers.

3. Understand a fraction $\frac{a}{b}$ with $a > 1$ as a sum of fractions $\frac{1}{b}$.

    b. Decompose a fraction into a sum of fractions with the same denominator in more than one way, recording each decomposition by an equation. Justify decompositions, e.g., by using a visual fraction model. *Examples:* $\frac{3}{8} = \frac{1}{8} + \frac{1}{8} + \frac{1}{8}; \frac{3}{8} = \frac{1}{8} + \frac{2}{8}; 2\frac{1}{8} = 1 + 1 + \frac{1}{8} = \frac{8}{8} + \frac{8}{8} + \frac{1}{8}$.

Fig. 6.12. Addition and subtraction of fractions, CCSSM 4.NF.3b
(NGA Center and CCSSO 2010, p. 30)

Summarizing Pedagogical Content Knowledge to Support Big Idea 4

Teaching the mathematical ideas in this chapter requires specialized knowledge related to the four components presented in the Introduction: learners, curriculum, instructional strategies, and assessment. The four sections that follow summarize some examples of these specialized knowledge bases in relation to Big Idea 4. Although we separate them to highlight their importance, we also recognize that they are connected and support one another.

Knowledge of learners

Some students mistakenly add the numerators and the denominators when adding fractions. This is an important misconception to anticipate and address. Helping students understand the meaning of the numerator, the denominator, and the unit, as well as the relationships among them, is critical for addressing this misconception.

Further, you may encounter students who use arguments similar to those shown in figure 6.11. Determining which students pay attention to sizes of pieces, such as student A, or which students lose track of what the unit is in particular problems, such as student B, is important.

Knowledge of curriculum

Mack (2004) suggests a variation on the traditional sequence of fraction problems. What sequence is used in the curricular materials at your grade level? What sequence is used at the grade levels before and after your grade level? Are problems posed in a context? Do the problems with different denominators start with problems in which one denominator is a multiple of another? Which models (area, discrete, or linear) do your curricular materials use? Do your students have opportunities to select and use different models as they add and subtract fractions? How are units identified in the problems? Are the contexts of the problems varied?

Knowledge of instructional strategies

As discussed on page 91, Mack (2004) posed the problem, "You have three-eighths of a medium pepperoni pizza. I give you two-eighths more of a medium pepperoni pizza. How much of a medium pepperoni pizza do you have now?" Note the strategy that she used to emphasize the unit. Rather than use numerical symbols ($\frac{3}{8}$ and $\frac{2}{8}$) in the problem, she spelled out the unit—"eighths." Often, students encounter fractions only as numerical symbols, rather than as words. Spelling out the names can be an effective instructional strategy when it is connected with a sustained emphasis on the meaning of units—especially for English language learners.

In figure 6.2, the problems involved $\frac{1}{2}$ and $\frac{1}{3}$ but, with the exception of problem 1, did not require or model the addition of $\frac{1}{2}$ and $\frac{1}{3}$. Using a variety of contextual problems is an instructional strategy that can provide your students with opportunities to read the problems and make sense of them.

Young children spend a significant amount of time counting, including counting all, skip counting, counting on, and counting back. Unfortunately, too often students in grades 3–5 do not spend as much time counting fractions. Such counting can be valuable. Try, for example, asking your students to count by halves starting at 0, count by thirds starting at 10, or count by halves starting at $\frac{1}{4}$. This strategy can help children see patterns and develop proficiencies that will strengthen their understandings of addition and subtraction.

Knowledge of assessment

Students use a variety of strategies to solve problems involving the addition or subtraction of fractions. Monitoring which students use direct modeling, derived facts, counting strategies, or representations to solve problems will help you facilitate their development of essential understandings. Because students' approaches will differ, it is important to assess their understanding by using problems from the four basic classes identified by Carpenter, Fennema, and Franke (1996). For example, you might use a joining problem, such as the following:

> Maria had $\frac{1}{2}$ cup of sugar in a measuring cup. She poured another $\frac{1}{3}$ cup of sugar into the same measuring cup. How much sugar is now in Maria's measuring cup?

Or you might offer a comparison problem, such as the one below:

> Craig has $\frac{3}{4}$ cup of sugar. Lauren has $\frac{6}{9}$ cup of sugar. Who has more sugar? How much more does that person have?

Assessing how students approach different classes of problems will provide you with information to help students continue to build their understanding of addition and subtraction of fractions.

Conclusion

The ideas and student work in this chapter have illustrated the complexity and challenges that you face in teaching students to compute with fractions with understanding. The preceding chapters have emphasized that to help students in grades 3–5 develop essential understandings related to fractions, you must give time and consideration to important ideas, such as the concept of unit, different interpretations of fractions, and notions of equivalency and comparison of fractions, before focusing on computation with fractions. We cannot minimize the complexity of integrating all these components or reduce it to a few superficial lessons about fractions. Like Barnett-Clarke and colleagues (2010), we have demonstrated the need to consider how these essential ideas are related to one another and how they support students' mathematical development as they progress to the middle and secondary levels. The next chapter highlights the alignment of the essential understandings discussed in the preceding chapters with mathematical concepts and topics that students encounter at lower and higher levels of learning.

practice

Chapter 7
Looking Back and Looking Ahead with Fractions

This chapter highlights how the essential understandings discussed in Chapters 1–6 align with ideas that students develop before and after grades 3–5. If students in grades 3–5 have gaps in their knowledge, you will need to assess their understanding of the foundational ideas that students are expected to develop in kindergarten through grade 2. We highlight these ideas in this chapter's first section. In the second section, we round out the chapter with a discussion of how the essential understandings presented in Chapters 1–6 connect with mathematics that students learn beyond grade 5. This discussion demonstrates how important it is in grades 3–5 for students to develop a deep understanding of the essential concepts that serve as a foundation for subsequent learning.

Supporting Knowledge in K–Grade 2 for Fractions in Grades 3–5

Teachers in kindergarten through grade 2 can support the development of students' essential understandings of fractions in grades 3–5. For example, in these years, it is important for teachers to help students establish a foundational understanding of the idea of a unit and focus on multiplicative thinking, while also learning to use various representations in ways that they can build on in grades 3–5. It is useful to discuss these aspects of students' foundational understanding in greater detail.

Focusing students on the unit

As students learn to count and measure, it is important to emphasize the unit that they are using and to help them recognize that the count or measure changes, depending on the unit that is established. As a simple example, students should recognize that if they hold up their hands, they can count the number of hands or the number of fingers, but the count will vary, depending on the unit (see fig. 7.1). Furthermore, when students play games such as concentration, they create pairs of cards that are "matches." When all the matches have been made and the game ends, they determine the winner by counting either the number of pairs or the number of cards (see fig. 7.2). For their determination to be correct, they need to be consistent in their use of the unit that they establish.

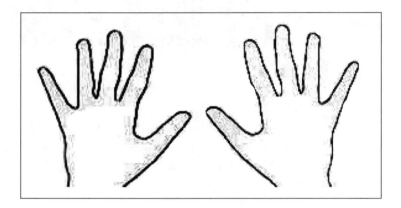

Fig. 7.1. Diagram of 2 hands or 10 fingers, depending on the unit chosen

Clarifying the unit that is counted is a key component of the counting process. Often the unit is assumed, though it is important to emphasize that typically when we count, we determine a quantity on the basis of some identified unit. Counting situations arise in a variety of contexts, and discussions occur quite naturally about one count or measure being more than, less than, or the same as another. These activities provide opportunities to introduce young students to the use of the language needed for comparing fractions in grades 3–5.

As students engage in situations that involve multiplication and place value, they must look at units in a new way, often referred to as *unitizing*. Unitizing means viewing objects or quantities in different-sized chunks (Lamon 1999). For example, unitizing would involve viewing the diagram in figure 7.3 as 30 blocks or as 3 groups of 10 blocks. Similarly, students would recognize 5 groups of 4 objects also as 20 objects. Depending on the unit or group, they could count the number of groups or the total number of objects.

Fig. 7.2. A model of 2 pairs or 4 cards in the game concentration

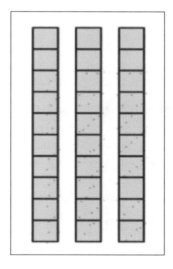

Fig. 7.3. Three groups of 10 blocks or 30 blocks

Measurement often provides a natural connection with and need for establishing a unit. Depending on what we are measuring, we need to select an appropriate unit, and we reason on the basis of that unit. For example, we have various units to choose from when measuring length, including inches, feet, miles, meters, kilometers, and light-years. We could also choose to measure the area, temperature, brightness, or humidity of our classroom. As we consider what attribute to measure, we need to select a unit that is appropriate for the attribute. Measuring the length of the classroom in shoe-lengths will result in a different count from measuring its length in paper clips. Discussions about the importance of the selection of a unit can help students understand the relationship between the unit and the count, providing a bridge to multiplicative reasoning and fractions.

Multiplicative thinking

Students in kindergarten through grade 2 often focus on reasoning in an additive manner. For example, when given 5 marbles, they consider how many more marbles they need to have 12 marbles. However, students in these early grades can also engage in multiplicative thinking, particularly in measurement situations in which the unit changes, but also in situations that involve equal-sized groups of objects.

The Common Core State Standards for Mathematics (CCSSM; National Governors Association Center for Best Practices and Council of Chief State School Officers [NGA Center and CCSSO] 2010) emphasize the importance of helping students in grade 2 lay the groundwork for multiplicative thinking. Figure 7.4 shows CCSSM's expectations that second graders will work with groups, pairing objects or counting them by 2s, and will work with rectangular arrays.

Common Core State Standards: Grade 2

Work with equal groups of objects to gain foundations for multiplication.

3. Determine whether a group of objects (up to 20) has an odd or even number of members, e.g., by pairing objects or counting them by 2s; write an equation to express an even number as a sum of two equal addends.

4. Use addition to find the total number of objects arranged in rectangular arrays with up to 5 rows and up to 5 columns; write an equation to express the total as a sum of equal addends.

Fig. 7.4. Operations and Algebraic Thinking, CCSSM 2.0A.3 and 4
(NGA Center and CCSSO 2010, p. 19)

It is important that we distinguish between multiplicative thinking and the process of multiplication. Confrey and Harel (1994) explain that reasoning mathematically about situations requires reasoning about *things* and *relationships*. It is important that students reason about things and relationships to develop and deepen their use of multiplicative thinking. Jacob and Willis (2001) emphasize that children must first come to recognize that multiplicative situations involve three aspects: groups of equal size, a number of groups, and a total amount. Figure 7.5 provides a representation of these three aspects that students must recognize: (a) groups that are equal in size (in this case, 4 circles), (b) the number of groups (in this case, 3 groups), and (c) the total number of circles (in this case, 12 circles). Furthermore, when children construct and coordinate these three factors in problem situations, they are thinking multiplicatively. Our desire is for students to recognize when and when not to reason multiplicatively, given a particular situation.

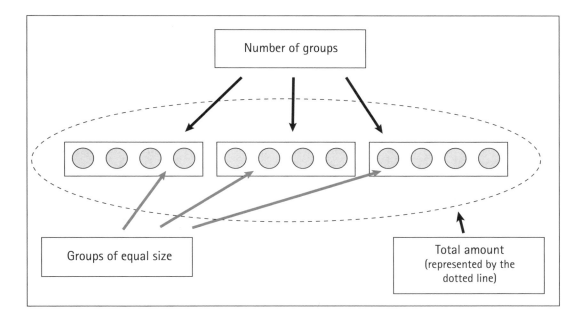

Fig. 7.5. Three aspects of a multiplicative situation

Jacob and Willis also argue that teachers need to recognize the difference between additive and multiplicative thinking:

> Multiplication is more than repeated addition, however, and its learning more complicated. While repeated addition may be an appropriate beginning, to maintain that interpretation of multiplication is ultimately disabling because it does not provide children with important multiplicative structures. (Jacob and Willis 2001, p. 306)

These words of caution about interpreting multiplication as repeated addition emphasize an important idea about the meaning established for multiplication. Students should recognize that 3×5 means 3 equal-sized groups of 5, rather than the often-introduced meaning of 3×5 as 5 added 3 times (that is, $5 + 5 + 5$). As students move toward developing an understanding of the meaning of fractions and of multiplication of fractions, they will discover that finding $\frac{2}{3}$ of a group of $\frac{1}{2}$ (that is, $\frac{2}{3} \times \frac{1}{2}$) makes sense, but that adding $\frac{1}{2}$ two-thirds times does not. Viewing multiplication as a relationship between the number of groups and the number in each group allows for a smooth transition to the meaning of fractions and of fraction multiplication.

Consider the following problem:

> You have 20 marbles. You would like to put the marbles into 4 bags, with the same number of marbles in each bag. How many marbles should you put into each bag?

This problem requires students to partition 20 marbles into 4 equal-sized groups. Such problems provide young students with a foundation for the partitioning that they will encounter later when they are introduced to fractions. Battista (2012) stresses the point: "Before students can understand fractions, they must understand partitioning. To partition a whole is to divide it into equal portions, like dividing a pizza equally among four people" (p. 1).

The difficulty of sharing problems can vary, depending on the numbers involved, the model in use (for example, discrete or area), and the presence of physical manipulatives (Van de Walle 2007). According to Van de Walle, children's initial strategies for sharing typically involve halving, so problems involving 2, 4, or 8 sharers are a good place to start. For instance, students could demonstrate how to share 10 cookies evenly among 4 people to consider situations that involve parts of wholes.

Contexts that introduce students to the meaning of "doubling" and "halving" and extending to situations that involve increasing an amount by 3 times (tripling) or 4 times (quadrupling) encourage multiplicative reasoning. The following situation, for example, promotes multiplicative reasoning:

> Sarah has 8 toy cars. Martha has 2 times as many toy cars as Sarah. How many toy cars does Martha have?

Such situations promote *iterating* the group—that is, *creating duplicates* of a unit in a process similar to what students must do when they create $\frac{2}{8}$. The toy car problem refers to 2 groups of 8 objects, whereas a situation involving $\frac{2}{8}$ refers to 2 groups of $\frac{1}{8}$.

Both iterating and partitioning are critical processes that students use in developing an understanding of the meaning of fractions (Siebert and Gaskin 2006). Tasks that encourage students to partition groups of objects into equal-sized groups or to iterate groups of objects promote the use of multiplicative reasoning, which supports the learning of fractions.

Students can iterate and partition continuous objects as well as discrete ones. Consider the tasks in figures 7.6 and 7.7, which promote partitioning and iterating a portion of the number line.

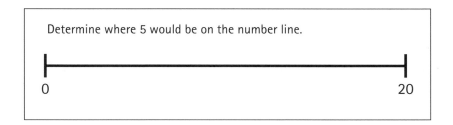

Fig. 7.6. A task involving placing 5 on the number line

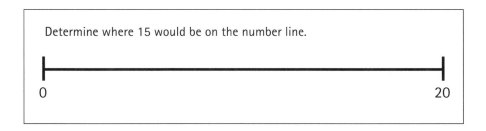

Fig. 7.7. A task involving placing 15 on the number line

Both number line tasks involve partitioning the number line by using a strategy of repeated halving. Further, these tasks require partitioning number lines of different lengths, an important skill that encourages students to apply multiplicative reasoning for partitioning rather than additive reasoning.

Contexts that involve measuring with different units also promote the use of multiplicative reasoning and support fraction development. Tasks such as those in figures 7.8 and 7.9, for example, motivate discussions about the relationships among units, the total quantity, and the measure or count.

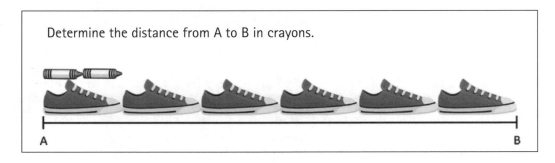

Fig. 7.8. A task involving determining the distance from A to B by using
the length of a shoe and the length of a crayon

Fig. 7.9. A task involving determining the distance from A to B by using
the length of a shoe and the length of a cupcake

Determining the number of crayons needed to measure the distance from A to B
in figure 7.8 can encourage the use of multiplicative reasoning, since the length of
2 crayons is the same as the length of each shoe. Therefore, students can reason
that the number of crayons needed to measure the distance from A to B is double
the number of shoes. The task in figure 7.9 encourages similar reasoning, since
the number of cupcakes needed to determine the length from A to B is 5 times the
number of shoes. Such multiplicative reasoning can lay a foundation for an understanding of units and multiplicative relationships that is similar to the meanings
that students establish for unit and non-unit fractions in grades 3–5.

Extending Fraction Knowledge in Grades 6–8

As we look ahead to students' development of their mathematical understanding of
ratio, rate, and percent in grades 6–8, we recognize the importance of developing
a deep understanding of the meaning of fractions in grades 3–5. In this section,
we briefly describe some of the important aspects of ratios, rates, and percent that

build on students' understanding of fractions. Figure 7.10 shows the CCSSM standards relating to ratios and proportional relationships for grade 6.

Common Core State Standards for Mathematics, Grade 6

Understand ratio concepts and use ratio reasoning to solve problems.

1. Understand the concept of a ratio and use ratio language to describe a ratio relationship between two quantities.

    For example, "The ratio of wings to beaks in the bird house at the zoo was 2:1, because for every 2 wings there was 1 beak." "For every vote candidate A received, candidate C received nearly three votes."

2. Understand the concept of a unit rate $\frac{a}{b}$ associated with a ratio $a{:}b$ with $b \neq 0$, and use rate language in the context of a ratio relationship.

    For example, "This recipe has a ratio of 3 cups of flour to 4 cups of sugar, so there is $\frac{3}{4}$ cup of flour for each cup of sugar." "We paid $75 for 15 hamburgers, which is a rate of $5 per hamburger."

3. Use ratio and rate reasoning to solve real-world and mathematical problems, e.g., by reasoning about tables of equivalent ratios, tape diagrams, double number line diagrams, or equations.

Fig. 7.10. Ratio and Proportional Relationships, CCSSM 6.RP.1–3
(NGA Center and CCSSO 2010, p. 42)

As noted by Lamon (1999), approximately half of the adult population demonstrates difficulty in reasoning proportionally, underscoring the challenges that we face in our efforts to develop in our students a deep understanding of ratio and the use of multiplicative reasoning. The discussion that follows emphasizes the importance of building on the underlying ideas in rational number to promote the use of proportional reasoning.

In this chapter, we use *ratio* to mean the comparison of two quantities, and we define *proportion* as an equivalence relationship between two ratios. As these definitions show, a proportion is a particular relationship—equivalence—between two ratios. Therefore, work with ratios and work with proportions require similar reasoning. A ratio involves recognizing a relationship as multiplicative rather than as additive. For example, consider the task in figure 7.11, which compares the number of cookies belonging to Michael and Seth.

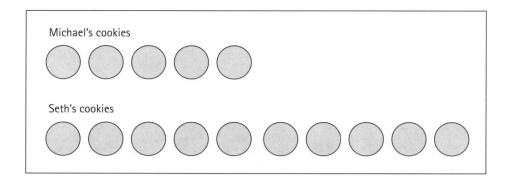

Fig. 7.11. Comparing the numbers of Michael's and Seth's cookies

We could compare the numbers of cookies that Michael and Seth have in two different ways. The first way would be additive: we could note that Seth has 5 more cookies than Michael, or that Michael has 5 fewer cookies than Seth. The second way would be multiplicative: we could note that Seth has twice as many cookies as Michael, or that Michael has $\frac{1}{2}$ as many cookies as Seth. This second comparison is how we would write a ratio to compare the numbers of cookies for Michael and Seth, since we could write the ratio of cookies for Michael to Seth as 1:2. Ratio comparisons are multiplicative comparisons that build on the multiplicative reasoning that students develop in working with fractions in grades 3–5.

We emphasized the need to refer to the same unit in Chapter 4, in our discussion of generating equivalent fractions and comparing fractions, and again in Chapter 6, in our discussion of adding and subtracting fractions. We believe that the importance of referring to the unit cannot be overstated. CCSSM emphasizes the reference to the unit in various standards related to fractions for grades 3–5. The unit is often overlooked despite its importance in students' developing understanding of rational numbers and skill in reasoning with rates, ratios, and percent. To complicate matters further, the notation used in the United States for fractions and ratios is the same (for example, we may use the symbols $\frac{3}{4}$ and $\frac{6}{8}$ to refer to the fractions three-fourths and six-eighths or the ratios of 3:4 and 6:8). This shared notation for fractions and ratios may cause difficulty for students in making sense of whether a fraction or a ratio is referred to in a particular context.

Students in grades 3–5 initially develop an understanding of the meaning of fractions by considering situations where they solve problems that involve a clearly defined unit or whole. This work with fractions in grades 3–5 lays a foundation that enables students in grades 6–8 to engage in reasoning that involves different units. Ratios, rates, and percent are often used to compare situations where the unit differs. Moreover, a deep understanding of rational number, ratios, percent, and rates requires the use of multiplicative, rather than additive, thinking. For example, suppose that we wanted to compare the data in the chart in figure 7.12 to determine which company sells the most reliable light bulbs.

	Number of defective light bulbs	Total number of light bulbs tested
Company A	10	120
Company B	22	500
Company C	7	50

Fig. 7.12. A chart showing the number of defective light bulbs for three companies

An additive comparison results in misleading conclusions in determining the company that sells the most reliable light bulbs. Students who reason additively, and not multiplicatively, would choose company C, despite the fact that this company has the highest percentage of defective light bulbs.

In grades 3–5, students build on their understanding of creating equal groups and partitioning groups into smaller groups of equal size as a foundation for the multiplicative understanding necessary for working with rational numbers and ratios. Students in these grades also investigate situations in which it is appropriate to use partitioning and create equal-sized groups and situations in which it is not. In grades 6–8, students extend this reasoning to contexts that involve a constant rate, such as situations that involve equal rates of speed, the cost of an item per particular unit, and scale models or drawings. Further, students learn to use ratios to convert measurement units and transform units.

One important goal is for students to recognize the underlying structure and consider the conditions that exist when using proportional reasoning. To investigate this further, solve the problems in figure 7.13. Think about the reasoning that you apply when approaching these problems.

Pen problem

The cost of 3 pens is $2.40. At the same price, how much will 10 pens cost?

Sprinting problem

Mary's best time for running 100 yards is 15 seconds. How long will it take Mary to run 500 yards?

Track problem

Sue and Julie were running at the same speed around a track. Sue started first. When she had run 9 laps, Julie had run 3 laps. When Julie completed 15 laps, how many laps had Sue run? (Cramer, Post, and Currier 1993)

Fig. 7.13. Three types of problems that students may encounter in grades 6–8

The problems in figure 7.13 have similarities in structure that suggest that proportional reasoning may be applied in all three. However, the relationships that exist in these situations differ, and in fact the Pen problem presents the only situation in which we can directly apply proportional reasoning. This problem states that we are to determine the cost of 10 pens "at the same price," given the cost of 3 pens. Therefore, the relationship between pens and cost remains the same. In the case of the Sprinting problem, it would be incorrect to assume that Mary can run 500 yards at the same rate that she runs 100 yards. Therefore, reasoning proportionally would provide an underestimate of Mary's time for running 500 yards, since we would expect her to run more slowly than she would for 100 yards.

The Track problem is likely to mislead both adults and students into applying proportional reasoning. Many would assume that the initial relationship between Sue and Julie, with Sue having run 3 times as many laps as Julie, continued. Thus, they would conclude that when Julie completed 15 laps, Sue had run 45. However, the problem states that Sue and Julie were running at the same rate. Therefore, Sue would maintain her 6-lap advantage over Julie as long as they continued running. When this problem was given to 33 preservice elementary teachers, almost all of them applied proportional reasoning and concluded that Sue had run 45 laps. Researchers have argued that such misunderstanding points to the superficial reasoning that many students develop in their school experience. It is important that students recognize when and when not to apply proportional reasoning. They should carefully consider the extent to which the initial relationship among quantities can be extended to other situations.

Prior understanding of equivalent fractions and of comparing rational numbers builds a foundation for using proportions and comparing ratios. For example, students in grades 3–5 learn to compare fractions such as $\frac{3}{4}$ and $\frac{8}{10}$ and decimals such as 0.35 and 0.268. These comparisons assist students when they compare ratios, percentages, and rates, since these quantities are often expressed as fractions and decimals. Therefore, students can apply their initial understanding of rational numbers to these new situations.

Ratios are similar to rational numbers in that some ratios are part-whole comparisons. A percentage is a ratio that is typically used to compare the part to a whole. When students work with rational numbers, they apply part-whole relationships. For example, students may consider a representation that could be used to designate the portion of students in the school who are involved in band. This amount could be represented as a fraction, a decimal, or a percentage. Each of these could be considered as a part-whole ratio that represents this relationship.

However, ratios can also be used to compare part-part relationships, and to compare other relationships that are neither part-part nor part-whole. For an example of a part-part ratio, consider the ratio of band students to non-band students in the school. Note that this relationship is a ratio, but it is not a rational number, since the relationship is not part-whole. Furthermore, ratios such as the number of miles per hours traveled, number of centimeters per inch, and cost per candy bar are rates that describe relationships but are neither part-whole nor part-part ratios. As noted previously, the many uses and representations of ratios can be confusing because of the different referents and the similarities in representing these ratios to the use of rational numbers. CCSSM expects students in grade 7 to reason about ratios in more sophisticated ways than students in grade 6, as shown in figure 7.14.

Common Core State Standards for Mathematics, Grade 7

Analyze proportional relationships and use them to solve real-world and mathematical problems.

3.   Use proportional relationships to solve multistep ratio and percent problems.

   Examples: simple interest, tax, markups and markdowns, gratuities and commissions, fees, percent increase and decrease, percent error.

Fig. 7.14. Ratio and Proportional Relationships, CCSSM 7.RP.3
(NGA Center and CCSSO 2010, p. 48)

Students in grades 6–8 learn about situations involving discounts as well as percent increase and decrease. The ideas discussed in Chapter 5 are critical to helping students develop these essential understandings in later grades. To explore the impact of the unit further, solve the problems in figures 7.15 and 7.16. After you solve them, consider on your own or discuss with others the impact of the unit in these situations.

Juanita is trying to determine the best price for the clothes that she plans to purchase at store A, store B, or store C. All three discount stores list the same retail prices for the clothes, but each one advertises a different discount:

Discount Store A:
30% off the original price, and then another 20% taken off the sale price

Discount Store B:
20% off the original price, and then another 30% taken off the sale price

Discount Store C:
50% off the original price

Which store is offering the best purchase price? How do the different discounts on the clothing affect your decision about which store is offering the best price?

Fig. 7.15. Best Deal problem

Samuel is wondering about the change in the value of a mutual fund that he owns. He is unsure how the current value of his mutual fund compares with its initial value. At the end of the first day, Samuel's mutual fund showed a 10% decrease in value. At the end of the second day, Samuel's mutual fund increased 10% from its value at the end of day 1. At the end of day 2, is the value of Samuel's mutual fund more than its original value, less than its original value, or the same as its original value? How would changing the original value of Samuel's mutual fund affect your response?

Fig. 7.16. Mutual Fund problem

Both the Best Deal problem and the Mutual Fund problem can bring to the surface important areas of confusion for students. In the case of the Best Deal problem, many are likely to believe that all three discount stores are offering the same "best price." However, this reasoning is incorrect (unless the sale price is $0). At stores A and B, the 30% and 20% discounts refer to different units. One of the discounts refers to the original price, and the other refers to the sale price after the first discount has been applied to the original price. At store A, the 30% discount refers to the original price, and the 20% discount refers to the sale price after the 30% has been discounted. At store B, the situation is reversed: the 20% discount refers to the original price, and the 30% discount refers to the sale price after the 20% has been discounted. At both stores, because the sale price after the first discount is a smaller unit than the original price, the result is not equivalent to taking 50% off the original price (or 20% off the original price and 30% off the original price). Thus, store C provides the best deal for any original price that is a positive price.

A second area of confusion for students in the Best Deal problem relates to the question of whether store A or store B provides the better deal. Many are surprised to find that both discounts result in the same final price, no matter what the price. Again, the reference to the unit is important.

Draw an area model displaying a 10 × 10 rectangle, and consider how removing 30% of the total area would look. Then consider how removing 20% of the remaining area would look. Draw a second 10 × 10 rectangle, and remove 20% of the total area, and then remove 30% of the remaining area. You might also note that for store A, you could multiply the original price by 0.7 to determine the price after the initial discount and then multiply that result by 0.8 to determine the final price. For store B, you could multiply the original price by 0.8 to determine the price after the initial discount and then multiply that result by 0.7 to determine the final price. Because you are simply multiplying the original price by the same two quantities, the commutative property ensures that your result will be the same in both cases.

The Mutual Fund problem causes similar difficulties for students who lack an awareness of the importance of the unit. The decrease of 10% by the end of day 1 refers to the original value of the mutual fund, and the increase of 10% by the end of day 2 refers to the value of the mutual fund at the end of day 1. Because these values are different, the original value of the mutual fund is more than the value of the fund at the end of day 2 for any positive original value of the fund.

Both the Mutual Fund problem and the Best Deal problem demonstrate the importance of emphasizing the reference to the unit in grades 3–5. For students to gain a deep understanding of percentage, they must recognize that they need to attend to the unit carefully in these situations.

Conclusion

Work with fractions in grades 3–5 provides an entry point for reasoning about various mathematical topics such as ratio, rate, proportion, and percentage. Moreover, this work influences student reasoning in many mathematical strands, including geometry, probability, statistics, and algebra. In geometry, students examine scale factors, scale drawings, and the ratio of the areas of figures when the ratio of the side lengths changes. In probability and statistics, students compare probabilities that are displayed as ratios, and they consider rational numbers that represent the portion of standard deviation from the mean for normal distributions. In algebra and calculus, students study rational expressions and consider their behavior when the numerator or the denominator increases or decreases in value. Students' development of a strong foundation in rational number, including understanding the relationship of a rational number to the unit and the meaning of the numerator and the denominator, is critical for their later work. Considerable time and effort are needed in grades 3–5 to ensure that students develop a deep understanding that will enable them to be successful as they move to grades 6–8.

Appendix 1
The Big Ideas and Essential Understandings for Rational Numbers

This book focuses on essential understandings that are identified and discussed in *Developing Essential Understanding of Rational Numbers for Teaching Mathematics in Grades 3–5* (Barnett-Clarke et al. 2010). For the reader's convenience, the full list of the big ideas and essential understandings in that book is reproduced below. The big idea and essential understandings that are the special focus of this book are highlighted in orange.

Big Idea 1. Extending from whole numbers to rational numbers creates a more powerful and complicated number system.

Essential Understanding 1*a*. Rational numbers are a natural extension of the way that we use numbers.

Essential Understanding 1*b*. The rational numbers are a set of numbers that includes the whole numbers and integers as well as numbers that can be written as the quotient of two integers, $a \div b$, where b is not zero.

Essential Understanding 1*c*. The rational numbers allow us to solve problems that are not possible to solve with just whole numbers or integers.

Big Idea 2. Rational numbers have multiple interpretations, and making sense of them depends on identifying the unit.

Essential Understanding 2*a*. The concept of *unit* is fundamental to the interpretation of rational numbers.

Essential Understanding 2*b*. One interpretation of a rational number is as a part-whole relationship.

Essential Understanding 2*c*. One interpretation of a rational number is as a measure.

Essential Understanding 2*d*. One interpretation of a rational number is as a quotient.

Essential Understanding 2*e*. One interpretation of a rational number is as a ratio.

Essential Understanding 2*f*. One interpretation of a rational number is as an operator.

Essential Understanding 2*g*. Whole number conceptions of *unit* become more complex when extended to rational numbers.

Big Idea 3. Any rational number can be represented in infinitely many equivalent symbolic forms.

Essential Understanding 3*a*. Any rational number can be expressed as a fraction in an infinite number of ways.

Essential Understanding 3*b*. Between any two rational numbers there are infinitely many rational numbers.

Essential Understanding 3*c*. A rational number can be expressed as a decimal.

Big Idea 4. Computation with rational numbers is an extension of computation with whole numbers but introduces some new ideas and processes.

Essential Understanding 4*a*. The interpretations of the operations on rational numbers are essentially the same as those on whole numbers, but some interpretations require adaptation, and the algorithms are different.

Essential Understanding 4*b*. Estimation and mental math are more complex with rational numbers than with whole numbers.

Appendix 2
Resources for Teachers

The following list highlights a few of the many books, articles, videos, and websites that are helpful resources for teaching fractions in grades 3–5. Abstracts from the publishers provide brief descriptions of the resources.

Books

Battista, Michael T. *Cognition-Based Assessment & Teaching of Fractions: Building on Students' Reasoning.* Portsmouth, N.H.: Heinemann, 2012.

> This book introduces a research-based framework that describes the development of students' thinking and learning in relation to fractions. The author discusses how teachers can build on students' reasoning during instruction.

Empson, Susan, and Linda Levi. *Extending Children's Mathematics: Fractions & Decimals— Innovations in Cognitively Guided Instruction.* Portsmouth, N.H.: Heinemann, 2011.

> The authors provide important insights into children's thinking and alternative approaches to solving problems through the use of student work, classroom vignettes, "Teacher Commentaries" from the field, sample problems, and instructional guides. Three themes are highlighted throughout:
>
> - Building understanding of fractions and decimals by discussing and solving word problems
> - Advancing children's strategies for solving fraction word problems and equations from direct modeling through relational thinking
> - Designing instruction that capitalizes on students' relational thinking strategies to integrate algebra into the teaching and learning of fractions.

Fosnot, Catherine, and Maarten Dolk. *Young Mathematicians at Work: Constructing Fractions, Decimals, and Percents.* Portsmouth, N.H.: Heinemann, 2002.

> Focusing on the way in which children construct their knowledge of fractions, decimals, and percents, the authors—

- contrast word problems with true problematic situations that support and enhance investigation and inquiry;
- provide strategies to help teachers;
- explore the cultural and historical development of fractions, decimals, and their equivalents and the ways in which children develop similar ideas and strategies;
- define and give examples of modeling, noting the importance of context;
- discuss calculation using number sense and the role of algorithms in computation instruction; and
- describe how to strengthen performance and portfolio assessment.

Goldenstein, Donna, Babette Jackson, and Carne Barnett-Clarke. *Fractions, Decimals, Ratios, and Percents: Hard to Teach and Hard to Learn?* Portsmouth, N.H.: Heinemann, 1994.

The authors use a case discussion approach to promote dialogue among teachers. The cases are written by upper elementary and middle school teachers about their experiences in presenting fractions, decimals, ratios, and percents. The cases stimulate reflection and dialogue about problems encountered in teaching these concepts in the classroom. The accompanying Facilitator's Discussion Guide provides practical suggestions for initiating and leading case discussion groups.

Van de Walle, John, Karen Karp, and Jennifer Bay-Williams. *Elementary and Middle School Mathematics: Teaching Developmentally.* Needham Heights, Mass.: Allyn & Bacon, 2010.

The authors wrote this book to help teachers understand mathematics and become confident in their ability to teach the subject to children in kindergarten through eighth grade. The chapters related to the teaching and learning of fractions provide ideas and insights that will support teachers as they design and implement their lessons.

Articles

Bray, Wendy S., and Laura Abreu-Sanchez. "Using Number Sense to Compare Fractions." *Teaching Children Mathematics* 17 (September 2010): 90–97.

The authors, two third-grade classroom teachers, found that giving particular attention to the use of real-world contexts, mental imagery, and manipulatives brought students success as they moved from using models to problem solving and reasoning.

Chick, Candace, Cornelia Tierney, and Judy Storeygard. "Seeing Students' Knowledge of Fractions: Candace's Inclusive Classroom." *Teaching Children Mathematics* 14 (August 2007): 52–57.

A fifth-grade teacher whose inclusive classroom brings together different learning styles and kinds of understanding describes what she learned from observing two students as they solved traditional fraction problems by using a clock face.

Empson, Susan B. "Research into Practice: Using Sharing Situations to Help Children Learn Fractions." *Teaching Children Mathematics* 2 (October 1995): 110–14.

> The author demonstrates the powerful reasoning that children bring to the classroom through the use of sharing situations.

————. "Equal Sharing and the Roots of Fraction Equivalence." *Teaching Children Mathematics* 7 (March 2001): 421–25.

> The author discusses examples of children's invented equal-sharing strategies that lay a foundation for reasoning about equivalence by connecting ideas of multiplication, division, and fractions.

Mack, Nancy K. "Learning Fractions with Understanding: Building on Informal Knowledge." *Journal for Research in Mathematics Education* 21 (January 1990): 16–32.

> Eight sixth-grade students received individualized instruction on addition and subtraction of fractions in a one-on-one setting for six weeks. The instruction was designed to build on the students' informal knowledge of fractions. All the students possessed a rich store of such knowledge, based on partitioning units and treating the parts as whole numbers, but their informal knowledge was initially disconnected from their knowledge of fraction symbols and procedures. The students related fraction symbols and procedures to their informal knowledge in ways that were meaningful to them; however, the students' knowledge of rote procedures frequently interfered with their attempts to build on their informal knowledge.

————. "Connecting to Develop Computational Fluency with Fractions." *Teaching Children Mathematics* 11 (November 2004): 226–32.

> The author explains how intermediate-grades students developed computational fluency in adding and subtracting fractions by focusing on the idea of operating on like-sized units and looking for similarities between problems and solution strategies. Teachers can use these problems and solution strategies in the classroom to reach a variety of diverse learners.

Norton, Anderson. "What's on Your Nation's Report Card?" *Teaching Children Mathematics* 13 (February 2007): 315–19.

> The author reports on a sample of fourth graders' responses to the 2003 NAEP exam. Readers are invited to analyze the responses to assess students' concepts of fractions.

Shaughnessy, Megan M. "Identify Fractions and Decimals on a Number Line." *Teaching Children Mathematics* 17 (March 2011): 428–34.

> The author discusses students' understanding of fractions and decimals on the number line and analyzes their erroneous strategies.

Siebert, Daniel, and Nicole Gaskin. "Creating, Naming, and Justifying Fractions." *Teaching Children Mathematics* 12 (April 2006): 394–400.

> For students to develop meaningful conceptions of fractions and fraction operations, they need to think of fractions in other ways than just as whole-number combinations. The authors suggest two powerful images for thinking about fractions in

ways that move beyond whole-number reasoning and can be used in the classroom to help support the teaching of fractions.

Watanabe, Tad. "Representations in Teaching and Learning Fractions." *Teaching Children Mathematics* 8 (April 2002): 457–63.

> The author describes various representations, including length, area, and set models, as well as fraction notation and the use of fraction language.

Whitin, David J., and Phyllis Whitin. "Making Sense of Fractions and Percentages." *Teaching Children Mathematics* 18 (April 2012): 490–96.

> Constructing pie charts has a meaningful context for fifth graders engaged in a long-term study about commercial advertising during children's television programs.

Wilson, P. Holt, Marrielle Myers, Cyndi Edgington, and Jere Confrey. "Fair Shares, Matey, or Walk the Plank." *Teaching Children Mathematics* 18 (April 2012): 482–89.

> The authors discuss how to teach young children to create equal-sized groups to build a flexible, connected understanding of, and ability to reason about, ratios, fractions, and multiplicative operations.

Videos

Integrating Mathematics and Pedagogy (IMAP)
http://www.sci.sdsu.edu/CRMSE/IMAP/video.html

> *IMAP: Select Videos of Children's Reasoning* is a CD containing twenty-five video clips of elementary school children engaged in mathematical thinking. The CD runs on PC and Mac platforms and comes with an interface that includes the transcript (full or synchronized) and background information for each clip. Also included on the CD is a video guide containing questions to consider before and after viewing each video clip, interviews that teachers or prospective teachers can use when working with children, and other resources.

Online Resources

Illuminations Lessons
http://illuminations.nctm.org/Lessons.aspx

> A project of the National Council of Teachers of Mathematics, Illuminations is part of the Verizon Thinkfinity program. The Illuminations website offers hundreds of standards-based lessons. Select the types of lessons that you want, as well as the appropriate grade band, and click "Search."

Equivalent Fractions
https://play.google.com/store/apps/details?id=air.com.learningtoday.equivalentFractions

> Equivalent Fractions is an app developed by Illuminations. It is also available as an online activity along with many other free math resources for teachers and students at http://illuminations.nctm.org. Students create equivalent fractions by

dividing and shading squares or circles, and match each fraction with its location on the number line. They also check their work as well as use the table feature to capture results and look for patterns. A "Build Your Own" mode allows students to select a particular value.

Illustrative Mathematics Project
http://www.illustrativemathematics.org

Illustrative Mathematics provides guidance to states, assessment consortia, testing companies, and curriculum developers by illustrating the range and types of mathematical work that support implementation of the Common Core State Standards for Mathematics. One tool on this website is a growing collection of mathematical tasks that are organized by standard for each grade level and illustrate important features of the indicated standard(s). The tasks on the website are not meant to be considered in isolation. Taken together in sets, these tasks are intended to illustrate a particular standard. Eventually, the site will showcase sets of tasks for each standard that—

- illuminate the central meaning of the standard and also show connections with other standards;
- clarify what is familiar about the standard and what is new with the advent of the Common Core State Standards;
- include both teaching and assessment tasks; and
- reflect the full range of difficulty that the standard expects students to master.

Appendix 3
Tasks

This book examines rich tasks that have been used in the classroom to bring to the surface students' understandings and misunderstandings about fractions. These tasks are reproduced here, in the order in which they appear in the book, for the reader's personal or classroom use.

Task 1

Do the brownies below have $\frac{1}{2}$ of the brownie shaded?

Explain your thinking.

(a)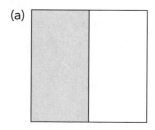

(b)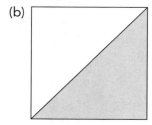

(c)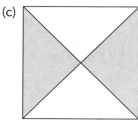

(d)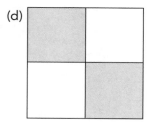

(e)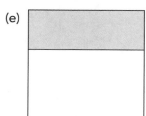

(f)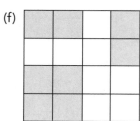

Task 2

Each picture below shows a brownie that is shared. You receive the shaded amount, and your friend receives the unshaded part.

Which of these brownies make fair shares? Explain your thinking.

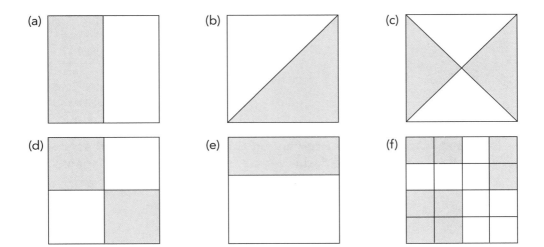

(a)

(b)

(c)

(d)

(e)

(f)

Task 3

(a) Each of the large squares shown below is a brownie. What fraction of the brownie is shaded in each case?

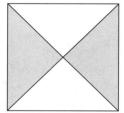 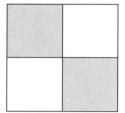 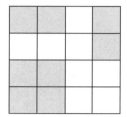

(b) What fraction of the brownie on the left is shaded? What fraction of the brownie is shaded if we cut the brownie in half again?

 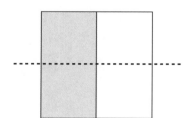

Task 4

Look at the shaded part of the two brownies below.
Circle the brownie that has a larger amount of a brownie
shaded.

 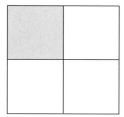

Explain why you think this is a larger amount of brownie.

Task 5

Read the thinking of the three students for the problem below.

1 brownie

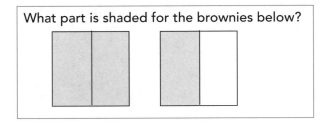

What part is shaded for the brownies below?

(a) Sally: I think $\frac{3}{4}$ of the 2 brownies is shaded. The brownies are cut into 4 equal parts and 3 are shaded.

Is Sally correct? Explain your thinking.

(b) Marcus: I think $1\frac{1}{2}$ brownies are shaded. One of the brownies is shaded and $\frac{1}{2}$ of another brownie is shaded, so $1\frac{1}{2}$ brownies are shaded.

Is Marcus correct? Explain your thinking.

(c) Demetrius: I think that $\frac{3}{2}$ of a brownie is shaded. Each brownie is cut in half and 3 of the halves are shaded.

Is Demetrius correct? Explain your thinking.

Task 6

Which of these has $\frac{3}{4}$ of the entire diagram shaded?

(a)

(b)

(c)

(d)
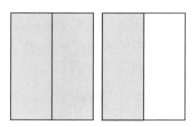

Task 7

(a) The shaded region below is $\frac{1}{4}$ of the whole. What does one whole look like?

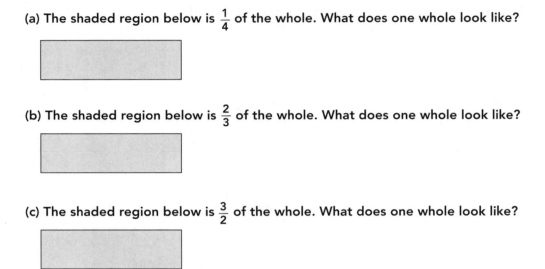

(b) The shaded region below is $\frac{2}{3}$ of the whole. What does one whole look like?

(c) The shaded region below is $\frac{3}{2}$ of the whole. What does one whole look like?

Task 8

What fraction is the upper rod if the lower rod is the whole?

$\dfrac{1}{2}$

$\dfrac{3}{4}$

Task 9

Read what Mary and Michelle say about the green triangle.

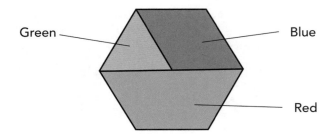

Green — Blue

Red

Mary: The green triangle represents $\frac{1}{3}$ of the pieces.

Michelle: The green triangle represents $\frac{1}{6}$ of the hexagon.

(a) Do you think Mary is correct? Explain your thinking.

(b) Do you think Michelle is correct? Explain your thinking.

Task 10

Ms. Jones asked her students to label the number line below. Write all the fractions that would be correct for where the question mark is on the number line.

Task 11

Read the responses of the students to the problem below.

List all fractions that could be located at the question mark below.

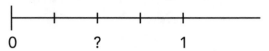

0   ?   1

(a) *Justin:* I think that this spot on the number line is $\frac{1}{2}$.
Is Justin correct? Explain your thinking.

(b) *Elizabeth:* I think that this spot on the number line is $\frac{2}{4}$.
Is Elizabeth correct? Explain your thinking.

(c) *Jane:* I think that this spot on the number line is $\frac{3}{5}$.
Is Jane correct? Explain your thinking.

(d) *Ethan:* I think that this spot on the number line is 0.5.
Is Ethan correct? Explain your thinking.

Task 12

Shade 0.6 of the grid.

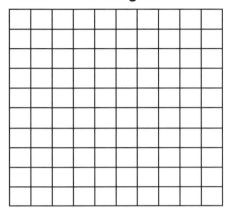

Write several fractions that could be used to represent the shaded portion.

Task 13

Shade 0.6 of the grid.

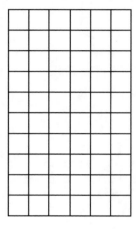

Write several fractions that could be used to represent the shaded portion.

References

Barnett-Clarke, Carne, William Fisher, Rick Marks, and Sharon Ross. *Developing Essential Understanding of Rational Numbers for Teaching Mathematics in Grades 3–5.* Essential Understanding Series. Reston, Va.: National Council of Teachers of Mathematics, 2010.

Bassarear, Tom. *Mathematics for Elementary School Teachers.* Boston: Houghton Mifflin, 1997.

Battista, Michael T. *Cognition-Based Assessment & Teaching of Fractions: Building on Students' Reasoning.* Portsmouth, N.H.: Heinemann, 2012.

Behr, Merlyn J., Richard Lesh, Thomas R. Post, and Edward A. Silver. "Rational Number Concepts." In *Acquisition of Mathematics Concepts and Processes*, edited by Richard A. Lesh and Marsha Landau, pp. 91–125. New York: Academic Press, 1983.

Bray, Wendy S., and Laura Abreu-Sanchez. "Using Number Sense to Compare Fractions." *Teaching Children Mathematics* 17 (September 2010): 90–97.

Caldwell, Janet H., Karen Karp, and Jennifer M. Bay-Williams. *Developing Essential Understanding of Addition and Subtraction for Teaching Mathematics in Prekindergarten–Grade 2.* Essential Understanding Series. Reston, Va.: National Council of Teachers of Mathematics, 2011.

Carpenter, Thomas P., Elizabeth Fennema, and Megan L. Franke. "A Knowledge Base for Reform in Primary Mathematics Instruction." *The Elementary School Journal* 97 (September 1996): 3–20.

Carraher, David William. "Some Relations among Fractions, Ratios, and Proportions." Paper presented at the Seventh International Congress on Mathematics Education (ICME-7), Quebec, Canada 1992.

———. "Learning about Fractions." In *Theories of Mathematical Learning*, edited by Lester P. Steffe, Pearla Nesher, Paul Cobb, Gerald A. Goldin, and Brian Greer, pp. 241–66. Mahwah, N.J.: Lawrence Erlbaum, 1996.

Chval, Kathryn B., and Óscar Chávez. "Designing Math Lessons for English Language Learners." *Mathematics Teaching in the Middle School* 17 (December 2011/January 2012): 261–65.

Chval, Kathryn B., Óscar Chávez, Sarah Pomerenke, and Kari Reams. "Enhancing Mathematics Lessons to Support All Students." In *Mathematics for Every Student: Responding to Diversity, Grades Pre-K–5*, edited by Dorothy Y. White and Julie Spitzer, pp. 43–52. Reston, Va.: National Council of Teachers of Mathematics, 2009.

Chval, Kathryn B., and Jane Davis. "The Gifted Student." *Mathematics Teaching in the Middle School* 14 (December 2008/January 2009): 267–74.

Clarke, Doug M., and Barbara A. Clarke. "Mathematics Teaching in Grades K–2: Painting a Picture of Challenging, Supportive, and Effective Classrooms." In *Perspectives on the Teaching of Mathematics*, Sixty-sixth Yearbook of the National Council of Teachers of Mathematics (NCTM), edited by Rheta N. Rubenstein, pp. 67–81. Reston, Va.: NCTM, 2004.

Confrey, Jere, and Guershon Harel. "Introduction." In *The Development of Multiplicative Reasoning in the Learning of Mathematics*, edited by Guershon Harel and Jere Confrey, pp. vii–xxviii. Albany: State University of New York Press, 1994.

Cramer, Kathleen, Thomas Post, and Sarah Currier. "Learning and Teaching Ratio and Proportions: Research Implications." In *Research Ideas for the Classroom, Middle Grades Mathematics*, edited by D. T. Owens, pp. 159–78. New York: Macmillan, 1993.

Dougherty, Barbara J. "Access to Algebra: A Process Approach." In *The Future of the Teaching and Learning of Algebra*, edited by Helen Chick, Kaye Stacey, Jill Vincent, and John Vincent, pp. 207–13. Victoria, Australia: University of Melbourne, 2001.

Empson, Susan B. "Research into Practice: Using Sharing Situations to Help Children Learn Fractions." *Teaching Children Mathematics* 2 (October 1995): 110–14.

———. "Organizing Diversity in Early Fraction Thinking." In *Making Sense of Fractions, Ratios, and Proportions*, 2002 Yearbook of the National Council of Teachers of Mathematics (NCTM), edited by Bonnie Litwiller, pp. 29–40. Reston, Va.: NCTM, 2002.

Fosnot, Catherine, and Maarten Dolk. *Young Mathematicians at Work: Constructing Fractions, Decimals, and Percents*. Portsmouth, N.H.: Heinemann, 2002.

Fuson, Karen C. "Research on Learning and Teaching Addition and Subtraction of Whole Numbers." In *Analysis of Arithmetic for Mathematics Teaching*, edited by Gaea Leinhardt, Ralph Putnam, and Rosemary A. Hattrup, pp. 53–187. Hillsdale, N.J.: Lawrence Erlbaum, 1992.

Grossman, Pamela. *The Making of a Teacher*. New York: Teachers College Press, 1990.

Gunderson, Agnes G., and Ethel Gunderson. "Fraction Concepts Held by Young Children." *The Arithmetic Teacher* 4 (October 1957): 168–73.

Hill, Heather C., Brian Rowan, and Deborah Loewenberg Ball. "Effects of Teachers' Mathematical Knowledge for Teaching on Student Achievement." *American Educational Research Journal* 42 (Summer 2005): 371–406.

Jacob, Lorraine, and Sue Willis. "Recognising the Difference between Additive and Multiplicative Thinking in Young Children." Paper presented at the 24th Annual MERGA Conference, Sydney, Australia, July, 2001.

Kabiri, Mary S., and Nancy Smith. "Turning Traditional Textbook Problems into Open-Ended Problems." *Mathematics Teaching in the Middle School* 9 (November 2003): 186–92.

Kamii, Constance, and Faye B. Clark. "Equivalent Fractions: Their Difficulty and Educational Implications." *The Journal of Mathematical Behavior* 14 (December 2005): 365–78.

Kieren, Thomas E. "Rational and Fractional Numbers as Mathematical and Personal Knowledge: Implications for Curriculum and Instruction." In *Analysis of Arithmetic for Mathematics Teaching*, edited by Gaea Leinhardt, Ralph Putnam, and Rosemary A. Hattrup, pp. 232–67. Hillsdale, N.J.: Lawrence Erlbaum, 1992.

Lamon, Susan J. *Teaching Fractions and Ratios for Understanding: Essential Content Knowledge and Instructional Strategies for Teachers.* Mahwah, N.J.: Lawrence Erlbaum, 1999.

———. "Rational Numbers and Proportional Reasoning: Toward a Theoretical Framework for Research." In *Second Handbook of Research on Mathematics Teaching and Learning*, edited by Frank K. Lester, Jr., pp. 629–68. Charlotte, N.C.: Information Age; Reston, Va.: National Council of Teachers of Mathematics, 2007.

Lesh, Richard, Marsha Landau, and Eric Hamilton. "Conceptual Models in Applied Problem Solving." In *Acquisition of Mathematics Concepts and Processes*, edited by Richard A. Lesh and Marsha Landau, pp. 263–343. New York: Academic Press, 1983.

Mack, Nancy K. "Learning Fractions with Understanding: Building on Informal Knowledge." *Journal for Research in Mathematics Education* 21 (January 1990): 16–32.

———. "Connecting to Develop Computational Fluency with Fractions." *Teaching Children Mathematics* 11 (November 2004): 226–32.

Magnusson, Shirley, Joseph Krajcik, and Hilda Borko. "Nature, Sources, and Development of Pedagogical Content Knowledge for Science Teaching." In *Examining Pedagogical Content Knowledge*, edited by Julie Gess-Newsome and Norman G. Lederman, pp. 95–132. Dordrecht, The Netherlands: Kluwer Academic, 1999.

Marks, Genée, and Judith Mousley. "Mathematics Education and Genre: Dare We Make the Process Writing Mistake Again?" *Language and Education* 4 (1990): 117–35.

National Governors Association Center for Best Practices and Council of Chief State School Officers (NGA Center and CCSSO). *Common Core State Standards for Mathematics. Common Core State Standards (College-and Career-Readiness Standards and K–12 Standards in English Language Arts and Math).* Washington, D.C.: NGA Center and CCSSO, 2010. http://www.corestandards.org.

National Research Council. *Adding It Up: Helping Children Learn Mathematics.* Mathematics Learning Study Committee, Jeremy Kilpatrick, Jane Swafford, and Bradford Findell, eds. Center for Education, Division of Behavioral and Social Sciences and Education. Washington, D.C.: National Academy Press, 2001.

Popham, W. James. "Defining and Enhancing Formative Assessment." Paper presented at the CCSSO State Collaborative on Assessment and Student Standards FAST meeting, Austin, Tex., October 10–13, 2006.

Pugalee, David K. "Connecting Writing to the Mathematics Curriculum." *Mathematics Teacher* 90 (1997): 308–10.

Reys, Barbara J., Ok-Kyeong Kim, and Jennifer M. Bay. "Establishing Fraction Benchmarks." *Mathematics Teaching in the Middle School* 4 (May 1999): 530–32.

Shaughnessy, Megan M. "Identify Fractions and Decimals on a Number Line." *Teaching Children Mathematics* 17 (March 2011): 428–34.

Shepard, Richard G. "Writing for Conceptual Development in Mathematics." *Journal of Mathematical Behavior* 12 (1993): 287–93.

Shulman, Lee S. "Those Who Understand: Knowledge Growth in Teaching." *Educational Researcher* 15, no. 2 (1986): 4–14.

————. "Knowledge and Teaching." *Harvard Educational Review* 57, no. 1 (1987): 1–22.

Siebert, Daniel, and Nicole Gaskin. "Creating, Naming, and Justifying Fractions." *Teaching Children Mathematics* 12 (April 2006): 394–400.

Silver, Edward A., Jeremy Kilpatrick, and Beth Schlesinger. *Thinking Through Mathematics: Fostering Inquiry and Communication in Mathematics Classrooms.* New York: College Board Publications, 1990.

Tall, David, and Shlomo Vinner. "Concept Image and Concept Definition in Mathematics with Particular Reference to Limits and Continuity." *Educational Studies in Mathematics* 12 (1981): 151–69.

Van de Walle, John A. *Elementary and Middle School Mathematics: Teaching Developmentally.* Boston: Pearson Education, 2007.

Watanabe, Tad. "Representations in Teaching and Learning Fractions." *Teaching Children Mathematics* 8 (April 2002): 457–63.

Wearne, Diana, and James Hiebert. "A Cognitive Approach to Meaningful Mathematics Instruction: Testing a Local Theory Using Decimal Numbers." *Journal for Research in Mathematics Education* 19 (November 1988): 371–84.

Widjaja, Wanty, Kaye Stacey, and Vicki Steinle. "Locating Negative Decimals on the Number Line: Insights into the Thinking of Pre-service Primary Teachers." *Journal of Mathematical Behavior* 30 (March 2011): 80–91.

Wiliam, Dylan. "Keeping Learning on Track: Classroom Assessment and the Regulation of Learning." In *Second Handbook of Research on Mathematics Teaching and Learning,* edited by Frank K. Lester, Jr., pp. 1053–1098. Charlotte, N.C.: Information Age; Reston, Va.: National Council of Teachers of Mathematics, 2007.

Wu, Hung-Hsi. "Teaching Fractions according to the Common Core Standards." http://math.berkeley.edu/~wu/CCSS-Fractions.pdf.

Yinger, Robert J. "The Conversation of Teaching: Patterns of Explanation in Mathematics Lessons." Paper presented at the meeting of the International Study Association on Teacher Thinking, Nottingham, England, May, 1998.